NATURE'S PERFECT FOOD

The story is told of an automaton constructed in such a way that it could play a winning game of chess, answering each move of an opponent with a countermove. A puppet in Turkish attire and with a hookah in its mouth sat before a chessboard placed on a large table. A system of mirrors created the illusion that this table was transparent from all sides. Actually, a little hunchback who was an expert chess player sat inside and guided the puppet's hand by means of strings. One can image a philosophical counterpart to this device. The puppet called "historical materialism" is to win all the time. It can easily be a match for anyone if it enlists the services of theology, which today, as we know, is wizened and has to keep out of sight.

—Walter Benjamin, "Theses on the Philosophy of History"

E. MELANIE DUPUIS

NATURE'S PERFECT FOOD

How Milk Became America's Drink

New York University Press • *New York and London*

NEW YORK UNIVERSITY PRESS
New York and London

Library of Congress Cataloging-in-Publication Data
DuPuis, E. Melanie (Erna Melanie) 1957–
Nature's perfect food : how milk became America's drink /
E. Melanie DuPuis
p. cm.
Includes bibliographical references (p.).
ISBN 0-8147-1937-6 (cloth : alk. paper)
ISBN 0-8147-1938-4 (paper : alk. paper)
1. Milk—History. 2. Milk—Social aspects. 3. Dairy industry—
United States—History. 4. Food habits—United States—History.
United States. I. Title.
GT2920.M55 D86 2001
641.3'71'0973—dc21 2001004677

New York University Press books are printed on acid-free paper,
and their binding materials are chosen for strength and durability.

Manufactured in the United States of America

10 9 8 7 6 5 4 3 2 1

Contents

Acknowledgments

A number of years ago, it became clear to me that I could spend the rest of my life writing this book. There remain many topics worth covering in more depth than I have in these pages. Milk, as many people who study it have told me, is a black hole that sucks you in and never let you escape. The real truth is that people who study milk never want to escape because the topic is endlessly fascinating. In the end, though, I realized that this book is as much about perfection as it is about food, and the book was done when I said what I had to say about American ideas of perfection using milk as a lens.

Using milk as a lens required, of course, reliance on that small but expert group of people whose profession is to think about and talk about milk. I have relied on these experts to help me understand what Americans have said and done to make this food over the last 150 years. For this I have relied heavily on Andrew Novakovic, James Pratt, and other dairy economists at the Cornell Department of Agricultural Economics (now the Department of Applied Economics and Management) in the interpretation of dairy efficiency studies and the untangling of the dairy regulatory system. While this book critiques the studies that came out of Cornell's Agricultural Economics Department, I certainly cannot criticize the careful attention given to me when I needed to clear up some confusion. In fact, my conversations with the earlier generation of agricultural economists at Cornell, particularly Bob Stanton and Harold Conklin, were what put me on the path of investigating milk from an environmental and political perspective. In California, Bees Butler helped me untangle that state's unusual dairy regulatory system. Harvey Jacobs and Jess Gilbert in Wisconsin sent me in some very valuable directions on the issue of land use planning and agriculture.

Of all my conversations with experts, the most fruitful were with Gould Colman, Cornell's University Archivist for many decades. He would answer my questions with a box, that is, a box in the archives I should look through to get my answers. He was the one who pointed me toward history, telling me that if I understood the history, I would understand how the human hand had a role in creating the present.

I must also thank David Barnett, committee assistant for the Assembly Agriculture Committee for the New York state legislature. As an intern in his office, I was the involuntary audience for many vocal phone conversations during David's negotiation of dairy legislation. My conversations with David made me realize the depth, richness, and mystery of dairy regulation.

There were a number of sociologists involved in studying dairying at the Department of Rural Sociology at Cornell during my own work there. My early work with Chuck Geisler on rBGH made me realize that the relationship between technological change and economic development in the dairy industry was not cut and dry. My dissertation advisor, Tom Lyson, and I worked together for many months, seeking a framework that would help us articulate the vast differences in dairy practices carried out by New York state farmers. Another member of my dissertation committee, Susan Christopherson, was the first to stop short in the hall during one of our conversations on my dissertation and say, "But the most interesting question is why we drink this stuff in the first place."

In the area of temperance and social reform in the nineteenth century, a three-hour phone conversation with Harry Levine helped steer me into more balanced waters. The interactive reviews in the Web pages of H-Rural were invaluable sources for the various perspectives on agrarian populism. I am truly grateful for what that mailing list has taught me about rural history.

To Linda Layne and the Social Studies of Science Department at Rensselaer Polytechnic Institute, I owe the opportunity of spending a postdoctoral year investigating the history of milk drinking. The members of that department freely shared their thoughts and skills, providing me with the tools necessary to look beyond political economy. I owe another huge debt to David Goodman, Margaret Fitzsimmons, Bill Friedland, Patricia Allen, and the rest of the Food Systems Study Group at University of California, Santa Cruz, who patiently sat through and responded to presentations of early chapter drafts.

This study group continues to provide me with ongoing, lively conversation about the sociology of agriculture.

I must also thank my terrific research assistants, Jennifer Snedecker, Jeannette Simmonds, Michael Schneidereit, and Allison Fletcher, who I always kept sending to the library for more and who always came back with more. In writing this book I returned again and again to the written exegeses they supplied me with, covering the literature on breast-feeding, natural theology, nineteenth-century diet, and other topics. They became so well acquainted with the literature that they ended up gently arguing me out of certain conclusions and into other ones. Librarians Mildred Pechman and Arlene Shaner, of the New York Academy of Medicine, helped me comb through archives of milk ephemera.

I have subjected a number of other people to early drafts of this work. Michael Kaplan, Sally Fox, Wally Goldfrank, Lisa Bunin, Luin Goldring, John Bowles, Richard Smoley, and my husband, Carl Pechman, pointed out numerous expository problems. I have learned that the most valuable gift one can give a scholar is a critical reading, and they all gave this gift to me. Michael Kaplan also teased me mercilessly over several months, forcing me to abandon several unexciting titles for the book and come up with a good one. My collaborative work with Danny Block, comparing the New York and Chicago milksheds, provided ways to think about the interrelationship between local politics and the "laws" of spatial organization.

Stephen Magro, at New York University Press, has been a tireless source of support in the writing of this book. Stephen signed me on long before my manuscript was ready to hit the streets, and he diligently worked with me to develop the book's potential. We both believe in the importance of bringing complex sociological ideas to a broader audience, and he invested significant time, effort, and patience in helping me meet this goal. Also thanks to the copyeditor, Rosalie Morales Kearns.

I would also like to thank Carl and my children, Eli and Abbey, for making my home a happy one. Many acknowledgments end with apologies for the nights and weekends that the author spent away from the family while writing a book. I won't do this because, for the most part, I did not work nights and weekends. To some extent, the time it has taken to write this book reflects this fact. On the other hand, the desire to spend time with my family made me a very productive 9-to-5er.

The people most responsible for this book, however, are the farmers of Cattaraugus and Chautauqua Counties whom I interviewed for a Cornell survey in the spring of 1988. They let me into their homes, sat me at their kitchen tables, and told me to be skeptical of the experts. They showed me that the productivist story of dairying was not the only story. Their insights made me look in different places, and in different ways, for history.

PART I

CONSUMPTION

1
Why Milk?

DO WE NEED to drink milk? Could we do without it? Should we? For over a century, American nutrition authorities have heralded milk as "nature's perfect food," as "indispensable," as "the most complete food." These milk boosters have ranged from consumer activists, to government nutritionists, to the American Dairy Council and its ubiquitous milk mustache ads. This pro-milk ideology has a long history, but is it true?

Recently, in the newest social movements around food, milk has lost favor. Vegan anti-milk rhetoric, for example, portrays the dairy industry as cruel to animals and milk as bad for humans. Recently, books with titles like *Milk: The Deadly Poison* and *Don't Drink Your Milk* have portrayed milk as unhealthy for humans. Controversies over genetically engineered hormones to produce milk and questions about antibiotic residues have also prompted a number of consumers to question whether the milk they drink each day is truly good for them. A food that was once glorified is now coming increasingly under question.

The public listens to the arguments for and against milk and asks, "Which view is correct?" Inevitably, when I tell people I am writing a book about milk, they think I plan to advocate one or the other side of this question. To their great disappointment, I have to answer that my book does not come down for or against milk. Instead, the book is about where milk came from, what questions have been asked about it, and how these questions have been answered.

To do this, I will go back to the beginnings of fresh milk drinking as a major dietary practice in the United States. By examining the growth of the fresh milk commercial enterprise, we will be able to see clearly how we came to drink milk. We can also see what other possible dietary practices, dairy or otherwise, got passed up along the way. The development of the milk industry was a part of the politics and ideologies of its times, from mid-nineteenth-century Victorian ideals to the ideals of our postindustrial age. How we drink milk, and why, has as much to do with the social relationships we share, and the way we think about these relationships, as it does with providing the body with nutrients. Ultimately, the question of whether or not milk is a "healthful" food has to be answered in a way that acknowledges the social context in which milk is consumed. "Healthy food" is an ideal that varies in relation to particular places and times.

Intrinsic to the rise of milk as the "perfect food" is the idea of perfection itself. Ideas about perfection provide a key to understanding modern society. The modern story of the march of progress entails the march to a perfect world. The industrial form of production, the hierarchical form of managerial bureaucracy, and the economic idea of supply and demand meeting at a single point all imply that there is one, single, perfect way to make, organize, market, and consume today's commodities. Ideas about perfection also underlie the way we look at our bodies and the food we consume to sustain our existence. National dietary guidelines, bodily standards of weight and blood composition, as well as homogeneous media images of health and beauty, imply that there is one, single, perfect body form. For example, national dietary guidelines, powerfully illustrated by the food pyramid, privilege one cultural form of eating at the expense of other cultures and food practices. It portrays one "best" way to eat and one "best" way to look.

The story of milk as a perfect food provides an exceptional opportunity to examine this sort of storytelling. Looking at how milk became America's ideal can help us gain an understanding of the perfect story and how it took hold of American society. To some extent, telling the perfect story requires "boosters," people explicitly working on behalf of a particular vision of perfection. As we will see, some people have told the milk story with that specific purpose. However, the vast majority of people writing the story of milk are not boosters per se; they see themselves as objective scientists, or as working in the best interest of society as a whole. Yet even in these cases, historians, economists, and others writing the history of milk have implicitly taken as fact many parts of the perfect story. In fact, this is the perfect story's express purpose: to take a particular interest and dress it in the robe of universality.

The perfect story portrays drinking this food fresh from the cow as an ancient custom and America as a milk-drinking nation from the beginning. In fact, as we will see, fluid milk drinking was an extremely minor aspect of the human diet until modern times. Even the northern

European countries consumed the vast majority of their dairy products in preserved or fermented forms such as cheese, sour milk, and yogurt. Americans in some parts of the country, particularly in the Northeast, ate dairy products on a regular basis; the "family cow" was quite common, not just on farms but in town backyards.[1] But the limited amounts of milk she provided the family went mostly to the production of butter and cheese, which were the center of the American dairy diet, particularly butter. Fluid milk drinking was an afterthought, and much of it was probably the fermented buttermilk, or "clabbered milk," that was a by-product of butter production, and this milk was primarily used for cooking or fed to hogs.[2] In other words, from colonial times to the mid-nineteenth century, fresh milk was not a major American beverage.

Yet histories of milk commonly describe the rise of sweet milk provision to the cities as a way to preserve a major traditional American food habit. The story starts with the fact of the family cow and assumes that she primarily supplied the family with fresh milk. In these histories, the rise in milk consumption was inevitable; the demand for this food was already present in the cities, as urbanites recalled their rural past. From this perspective, explaining the rise of milk drinking only involves the story of how to bring this desired rural substance safely to the city. From this perspective, the story of milk is "a portrayal of man's humanity, and sometimes inhumanity to man, and as in the truly popular drama, of a righteousness born of knowledge, triumphant in the end."[3]

In fact, sweet, fresh cow's milk began its real life as an American food in the mid-nineteenth century, primarily as a breast milk substitute for infants and a beverage for weaned children. It began not on primeval farms, but in the burgeoning city; not with the rise of sanitation, but before sanitary production was possible. Milk drinking as a new food habit began in the mid-nineteenth century, when this food was still dangerous, even deadly, as it was produced at that time. One historian has referred to the nineteenth-century city milk supply as "white poison."[4]

Despite increasing evidence that the milk supplied under this system was deadly, city people increasingly supplied this milk to their infants and children. Some opponents of this system went so far as to attest of the milk that "out of 100 children fed with it, 49 die yearly."[5] As early as the 1840s, a prominent milk reformer claimed that three-quarters of all infants and children in New York City were being raised (or dying) on cow's milk. While this claim was probably an exaggeration, it does indicate that substantial numbers of children were drinking cow's milk at the time.[6]

Yet by the 1880s, milk drinking as a food habit had begun to spread to the populace as a whole. By the 1940s, the average American was drinking over a pint of milk a day, twice as much as her 1880s counterpart.[7] Milk had become a major staple of the American diet. For most of the twentieth century, the vast majority of Americans—children, adults, older people—drank milk in large quantities.

If, in fact, the history of milk is not simply a happy reunion with a longed-for past, if the consumption of this substance in its sweet, fresh form coincided with the rise of the city, then another story needs to be told. How did this milk-drinking habit happen, and why did milk become an indispensable American dietary staple?

To answer this question requires looking at milk from its beginnings, in the cities where the demand for it first began. We also have to look at the making of milk, from the farms on which it has been produced to all those forces working when you open a refrigerator and reach for that carton. In other words, the answer to the milk question needs to look at this food from production to consumption, from the birth of the industry in the mid-nineteenth century to the "Got Milk?" campaign, from the nitty-gritty details of milk production to the place of perfection in American social thought.

One of the most amazing facts I discovered in my study of milk consumption over the last century is that no one has asked these questions in a systematic way. What I have discovered, looking at what everyone has said about milk—and people have said volumes—is that each student of this subject tells one small part of the story, with very little attention to other parts. This fractured talk about milk blinds us to some of the most important questions, the "why milk?" and "how milk?" questions that can illuminate one of our most basic food habits. Somehow, in studies of the dairy industry, its regulation, its science, and even in more historical and sociological studies of this food, the why and how of milk tend to get missed.

However, these more narrow approaches are my building blocks as I try to tell the larger story about milk. For that reason, this book is totally—even madly—interdisciplinary, taking information from such widely diverse disciplines as agricultural economics, rural sociology, geography, business history, women's history, cultural studies, media studies, and dairy science. In the following pages, you will be introduced to a wide variety of religious and social reform movements, economic studies of dairy farm efficiencies, feminist studies of the American family, and American ideas of nature. We will jump from economics to politics to culture and back again.

To answer this question about milk also requires asking why this question has been so invisible for so long. My answer will be that the conventional stories about milk fit into conventional ideas about social change. The following chapters will closely examine those conventional stories, what I call the "progress" and "downfall" stories. Both of these stories tend to depend too much on the idea of history as an automatic process and not enough on how history is the product of cumulative choices people make. To ask new questions about the rise of milk drinking will require the formulation of a different story, one that includes people, and not just processes, in the explanation.

As just about any American can tell you, milk production is highly regulated by the government. But this was not always so. Even now, with deregulation a keyword in any political agenda, Democrat or Republican, liberal or conservative, milk has continued to be one of the most heavily regulated commodities around. How did milk attain this privileged position in American politics? The answer to this question and the answer to the question above—why do we drink so much milk?—are intrinsically intertwined.

This is not just a city story, however. It is also a story about the countryside, particularly about how the relationship between the country and the city changed over the course of American history. The story of perfection first arose in the rural religious movements of the Second Great Awakening, then moved to cities to eventually become the basis for urban social reform movements. Eventually, urban professionals brought this story back to the country, as a tool to reshape farm production to fit a new urban vision of perfection. In this process, food and agriculture experts arose as powerful actors, telling a new story about perfect farming. This new urban-based vision of the countryside portrayed the farmer as the efficient, industrial provider of cheap urban sustenance. As farmers became less powerful in American politics, they became less able to have a voice in the creation of this dominant story of American life, including country life. As a result, farmers became invisible as actors in the industrial story of perfection.

To answer the important questions about milk requires heavy use of historical resources. However, this is not a history—or even a social history—of milk drinking. In these pages I examine the rise of a particular food habit in American life, as a social practice and as a story about that practice. I focus on the stories about milk, the deliberations, the public rhetoric and debate that created milk as a new, modern food. I also focus on the social, political, and economic relationships around dairy production itself, and how these relationships were recognized, ignored, celebrated, or deplored in the public—increasingly urban—deliberations on milk. These deliberations about milk were motivated by the desire for social reform. Since the mid-nineteenth century, milk has been a part of the American social change agenda, from the temperance movement to the Progressive movement, the New Deal, and finally the current movement against genetically engineered foods.

When I tell people I am writing a book on milk, inevitably they respond, "So you're going to talk about recombinant bovine growth hormone (rBGH)." I have written about this controversial genetically engineered hormone currently used to boost a cow's milk production,[8] and this book originally came out of my interest in rBGH. Nevertheless, those readers looking for an exposé of rBGH in this book will be disappointed. I am not interested in decrying the advent of biotechnology. Instead, I am more interested in looking at the anti-rBGH movement as one in a long line of social movements around milk.

I can tell you the simple answer to the "why we drink milk" question: milk is more than a food, it is an embodiment of the politics of American identity over the last 150 years. Therefore, the process of showing why we drink milk will involve looking at milk in the context of the rise of industrial food and modern eating in American culture. The answer to the "why we drink milk" question becomes apparent only if we look at this phenomenon as a product of a particular social and political history.

Using milk to represent the process of food industrialization overstates the case, to some extent. Fluid milk as a food has one strange quality that makes it unlike any other industrial food: the history of milk, unlike many other commodities such as wheat, rice, or even orange juice, is not a history of the march to globalization. Fresh milk production has remained relatively local, although this local aspect is in itself a political phenomenon (having to do with a public unwillingness to take some or all of the water out of fresh milk in order to move it around in its concentrated or powdered form). Therefore, the social and political context, even today, revolves around "milksheds," the areas around cities that still provide a particular metropolitan area with a significant proportion of its fluid milk.

Most histories of the industrial food system focus on the increased movement and processing of food, how food has been made fast and global. Bill Friedland, in his study of the California lettuce industry, notes the relationship between the introduction of fresh food into cities and the industrialization of the agricultural production process.[9] However, industrialization of lettuce production, like that of many other foods, involved a reorganization away from the local foodshed. Milk is one of the last foods to retain this intimate connection between freshness and local production, between farmer and consumer. In fact, the political and economic links between cities and their rural milksheds generally deepened with the rise of milk consumption.

Because fresh milk seems to be an exception to the processes of globalization, it is treated as a somewhat old-fashioned product, the remnant of a system kept in place by the power of the dairy lobby. In fact, drinking fresh, fluid milk from an unknown animal is an industrial form of consumption, even more today, when "fresh" milk is actually the recombined product of various dairy food components. In fact, the story of modern milk is the story of the modern corporation. As Alfred Chandler has noted, food corporations, including the major dairy corporations, represent some of the earliest firms organized along the lines of the modern industrial enterprise.[10] Providing this new food to cities involved some of the modern business strategies that characterize the global food structure today. For example, contemporary accounts of the global food system describe the multinational strategy of "global sourcing": playing one region of producers off another to lower the selling price. In fact, as we will see, such strategies were first used to control milk prices, and farmer agitation over prices, in milksheds over a century ago.

This continued localness of production also enables us to see more. Because both the production and consumption of this food has historically revolved around city milksheds, farmer-consumer relationships have tended to cover only a few political jurisdictions, thereby putting the politics and culture of this food into a clear light all the way from the farm to the table. In contrast, current global food politics often involve noncontiguous overlappings of regional, national, and global political institutions, making the relationships between farmers, food companies, consumers, and government officials harder to untangle.

Yet despite their eventual power to mold the tastes of the American public, the large industrial dairy corporations did not create the initial demand for milk. The recent Nestlé infant formula controversy, in which floods of corporate advertising convinced poor women in countries with inadequate and impure water supplies to replace breast milk with powdered formula, cannot be used to explain the drop in breastfeeding among urban women in the early nineteenth century. As I will show, fresh fluid milk and other breast milk substitutes were part of Americans' imaginations and desires even before a system adequate to provide these foods was in place.

The story of how milk became the "perfect food" tells us a great deal about how Americans have talked about food, thought about food, and thought about themselves in the last century. Therefore, while this book's primary subject is milk, in fact milk is a focus for a discussion of changes in American cultural relationships around food. Contemporary movies like *Soul Food* and *Babette's Feast* show how the dinner table, as a place for sharing food, signifies commonality and "collective consciousness" better than any other symbol. Yet that commonality often masks systems of repression, forcing people to "eat" particular notions of identity, purity, society, and nature. Milk was one of the first foods to become the subject of a public bargaining conversation over what "good" food was, and this debate over goodness overlay a larger but more hidden debate about the nature of American society. Therefore, while the rise of the milk industry required a reformulation of social conceptions of the body, particularly the bodies of women and children, these social conceptions were in flux before the industry arose to serve the new, socially created need.

THE PERFECTION OF MILK

Stories about perfection are in fact acts of power. In contexts where certain people have greater privilege than others, the social contingency of ideas tends to be replaced with one particular idea, a "perfect" idea of what society should pursue as a goal, and how people should live their lives. The classical philosophers of Athens provide a good example of such a situation. In a society controlled by free men, Plato's concepts about the imperfection of slaves and women tended to take hold. The perfect idea becomes a perfect story, which explains the past as a process in pursuit of the perfect idea—of health, happiness, or other desirable things. The social philosopher Walter Benjamin stated that this type of history required emptying the story of any other elements: "The concept of the historical progress of mankind cannot be sundered from the concept of its progression through a homogeneous, empty time." Exceptions get forgotten: "every image of the past that is not recognized by the present as one of its own concerns threatens to disappear irretrievably."[11]

From Plato onwards, elites have had the power to define the ideal form of living. Needless to say, these definitions have generally privileged elites themselves, giving them the right to speak, at the expense of others who are silenced. Plato's ideas about the perfect body and the right form of eating denigrated women and (often foreign) workers and privileged male Greek intellectuals. The privileged discourse about the perfection of milk has left out those people—mostly people of color—who are genetically lactose-intolerant. The perfect whiteness of this food and the white body genetically capable of digesting it in large quantities become linked. By declaring milk perfect, white northern Europeans announced their own perfection.

The story of milk, nature's perfect food, exemplifies this story of progress, perfection, and power. People tell this story as a way to explain the past, but they also tell the story as a way to determine the future. With the rise of modern industrial society, a new group of people—the professional elites—became America's official storytellers, those who told the country where it has been and where it is going. Two professional groups have told the official story of milk: historians and economists. Agricultural historians—often trained as economists—have told us about milk's past and agricultural economists have told us the story of its future.

To tell another story about milk first requires a critique of the official storytellers. However, this is not a critique of the disciplines of history and economics, which have more depth and breadth than illustrated here. Most historians and many economists today would argue with the modern story of perfection. Yet public discussions about milk have tended to be closed to those seeking to tell a less perfect story. In fact, until recently, the perfect story about milk received little challenge from any quarter. Only those with the professional credentials were invited to the table.

Yet today the voice of authority has come increasingly under question. Social movements addressing the environment, nuclear power, war, or a number of other issues have increasingly criticized the official story and the claims to authority of professional elites.[12] This "downfall of authority" has affected the discourse over food as well. I will review the rise of the professional elites who have told us the official story of milk and then examine the current popular challenge to their claims.

Perfectionist thinking is an act of power. It tells a story of historical change that makes social forces invisible. Instead, the motive propelling us along the rising graph is the unfolding of the perfect life. The perfect story can never be entirely contradicted because, in fact, the story has strong elements of truth. One cannot deny that milk drinking rose significantly over a hundred-year period in the United States. The reasons agricultural historians give for this rise—the improvement of the product, the education of the populace concerning its beneficial qualities—are substantially correct, *if one presumes that the product itself is perfect.* Only by questioning the perfection of milk can one look at the history of this product in a different way.

The idea of perfection creates a particular way of explaining historical change. If one way of life is perfect, then history is the process of moving toward or away from this perfect form. If there is one single goal, then there is one path toward that goal, one in which the idea of perfection—defined in terms such as bodily "health" or productive "efficiency"—is gradually achieved. This leads to two major stories: the "progress story" tells history as the march to perfection, while the "downfall story" tells history as the fall away from perfection.

Human downfall is the most common framework used to critique the story of perfection. Yet the downfall story is simply a mirror image of the story of perfection. The downfall story sees human society on a path toward perdition, as opposed to progress. Capitalism, industrialization, and/or state bureaucracies are taking us along this path. While the downfall story is concerned about relationships of power, it also, like the progress story, does not allow for a social explanation of change. Explanations of historical change instead follow a declensionist strategy, describing our present circumstances as the product of autonomous forces leading us to our downfall. The critique of these forces amounts to a rejection, but that rejection is simply a reaction. The critique depends on simply contradicting the progress story. The result is a rhetorical "white-is-black" effect: Everything that is true in the progress story is declared untrue in the story of social downfall.[13] In both cases, history is unilinear; only the direction changes. The idea of perfection and the concept of history as unilinear therefore go hand in hand.

The downfall story has become increasingly evident in the recent growth of the anti-milk movement and in the radical critique of industrial food in general. From this perspective, milk is not perfect, it is poison. As a rejection of the National Dairy Council's recommendations to drink several glasses of milk a day, the anti-dairy response is to drink no milk at all. In this movement, milk becomes associated with a wide variety of diseases and is denigrated as "pus." Dairy farmers are portrayed as cruel to animals, particularly calves who are denied their mother's milk.

There is a way of telling the milk story that critiques the progress story without simply becoming its Janus face. This third story, unlike the other two, brings to the fore social explanations, social choices, and events and practices that do not fit the trajectory. When we foreground social explanations, the relationships of power disguised by the other stories are made apparent. Parts of the story that have been hidden, ignored, or denigrated can be given their rightful place. Creating this third story requires seeing milk as neither perfection nor poison, but as a food that has been an intrinsic but problematic part of American society for over a century.

In the telling of this third story, the progress and downfall stories become keys. Identifying the rhetorical sleights of hand behind unilinear narratives provides a flashlight to look into the dark recesses where those things unexplainable by these narratives have been hidden. This is the process by which social evidence can be rediscovered and a third, more social, explanation can be forged.[14]

TELLING THE LESS THAN PERFECT STORY

How does one tell a less perfect, more open, more social story about milk? The open story cannot simply praise or blame modern social institutions. Instead, it seriously questions the basic assertions made by those representing the major institutions surrounding milk. What we will find is that milk production and consumption are not different parts of the story; we need to interweave them in order to understand how people created the food they considered perfect.

To tell this other story requires focusing on broader aspects of American society not commonly brought into the discussion. Religion, identity, and politics play key roles in this story. The rise of a powerful urban political voice becomes a major explanatory factor, laying the groundwork for the ways food is re-created in public discussion. The industrialization of the food system, from this perspective, is not an autonomous phenomenon but a product of certain political bargains based on certain dominant religious and cultural notions. The establishment of white racial hegemony and the celebration and purification of a white substance digested predominantly by this group become more than accidental.

Part 1 of this book concentrates on the rise of new notions of perfection and milk's role in that stream of American thought. These chapters are primarily about the city and the ways the city has rethought its vision of the countryside. In chapter 3, the discussion of issues surrounding the rise of the milk industry and the adulteration of milk pays attention to the changing identities of women—as mothers and as nurturing bodies—in the mid-nineteenth century. The examination of the early-twentieth-century controversy over pasteurization in chapter 4 will focus on mandatory pasteurization as an attempt by urban reformers to choose a particular economic structure for the industry as a way to perfect this food. Chapter 5 will show how efforts to envision the perfect child's body influenced the rise in milk drinking. Part 2 examines the effects of this new vision on dairy agriculture. Chapter 6 shows how the city and state agricultural officials forged an "industrial bargain" with larger farmers, re-creating the countryside around a new vision of perfection. Chapter 7 will describe how the urban revisioning of the countryside led to rural crisis: the milk strikes by farmers in the 1930s. Chapter 8 will examine how the New York state and federal governments went about resolving this crisis, through milk market control and a land use policy that labeled some farmers modern and some dispensable. Chapter 9 will examine the dairy policies of two other states, Wisconsin and California, to show that there were alternatives to the industrial bargain. In these states, governments emphasized adaptation to different resource and market environments, as opposed to New York state's triage policies.

With the final decline of the city as the center of American political life has come an accompanying fall in the voice of centralized authority.

The contemporary questioning of political voices of authority reflects this fall of the perfect story, as illustrated in the current questioning of the perfection of milk and the resistance to bovine growth hormone (rBGH) covered in chapter 10. Yet from this decline in the authoritarian voice comes the opportunity to create new forms of politics. I will examine these new forms of politics as they appear on today's milk cartons and how they "talk" to the consumer about rBGH.

The authoritarian, centralized voice of progress and its equally centralized downfall counterpart are basically "society-free" stories, attributing the development of this food to engines outside social influences. Separating milk from its social context, however, requires leaving out certain parts of the story, creating Benjamin's "homogeneous time." By tuning into the social bargaining conversation that has been part of the history of our food system, we can explain parts of milk's history that are commonly ignored in the other approaches. In this way we keep the milk story "open" to social explanations while allowing larger forces to also play a role. By allowing society back into the milk story, we can see how this product became part of our world through a series of human choices influenced as much by politics as by mechanisms commonly presented as neutral, such as science and the economy.

From this third approach, the heterogeneity of our nineteenth-century food system and the social and political reasons why one form of organization won out over others become clear. What also becomes clear is that our food system is the product of a social struggle over the definition of good food in the minds of consumers. A close look at the history of milk in the United States reveals this social struggle, the bargaining conversations, and the social choices that were made. These bargaining conversations have been played out in several arenas. Milk politics today is not simply the overt struggles over prices and safety but a more subtle "politics of representation" in which various actors in the food system lay claim to their visions through differing representations of food, nature, and society.[15]

The outcome of these social and economic bargaining processes has involved a coalescence of visions toward a dominant worldview concerning human-to-nature, human-to-food, and human-to-human relationships. Actors have created new dietary "accords" around changed definitions of nature and civilization, as well as new definitions of the relationships between the city and the country, gender and the body, society and nature.

This more complex explanation requires a new analysis of public battlegrounds, not as static struggles but as shifting arenas in which each victory, compromise, or capturing of prisoners redefines the footing of the actors.[16] I will describe this "footing" as the goals each actor has been fighting for, which involved shifting perceptions of goodness, perfection, and purity. Utopian ideals of perfect food have not been static, nor have they been the product simply of a hegemonically determined "false consciousness," and yet they have also been shaped in part by economic power structures. Reform goals were based on the perfectibility, or improvability, of each economic era, but that potential was the product of bargaining between actors, and that bargaining played upon commonly understood cultural relationships such as motherhood and nature, as they were defined during each period.

I will follow the history of the idea of perfection in America by focusing on that ultimate icon of perfection: milk, "nature's perfect food." In fact, the way people have talked about milk in America is so jarring in its metaphorical baggage, the ideas presented in its favor are so strange, and the people involved so passionate, that it provides us with an infinite series of interesting questions, a few of which are the focus of this book.

Finally, I would like to warn those who have picked up this book to get an answer to the question, "Is milk good for you?" that you will be disappointed. I will talk about the asking of this question—and the social context of this asking—but I will not give you an answer. By the end of this book, you will see why this question is not, and cannot be, answered.

2
The Perfect Food Story

THE VAST MAJORITY of people in the United States today consider milk an indispensable food. For the average American family, a refrigerator without milk is a compelling reason to run right out to the store. Milk remains, despite concerns over fat and cholesterol, a daily centerpiece of American nutrition. The USDA's food guide pyramid recommends two to three servings of dairy products a day, with the support of such professional groups as the American Dietetic Association. Public personalities—from Bill Clinton to Spike Lee—pose with "milk mustaches" in ubiquitous magazine advertisements.

The dairy boosterism of the milk mustache ads is not new. It reaches down through American history, long before the establishment of the National Dairy Council, long before scientific knowledge of vitamins and protein, much less of calcium. For more than a century, people have called milk a "perfect food." What is even more surprising than the long-standing nature of milk boosterism is the number of similarities between the earliest declarations of milk's perfection and more recent, more scientifically informed pronouncements on the topic. Although the reasoning has changed, the declaration of milk's perfection remains.

Milk's perfect story makes the history of its rise unproblematic. Perfect stories make the "why?" question unaskable, since perfection creates one story only: that of public enlightenment about and increased public desire for this food. The "perfect food" narrative describes milk's part in the rise of modern society, a perfectible society that could build a food system enabling people to have access to more, better food, safely and inexpensively. Milk is not only a better food, in this story, but it is an "essential" food, one necessary for optimal physical development. It is a history in which the perfection of the body through milk drinking and the perfection of society go hand in hand. Therefore, to ask the "why milk?" question means to challenge its "perfect food" narrative. In particular, it requires looking at how this story came to be told.

New York City, one of the earliest industrial mass markets for fluid milk, is an excellent place to focus in a history of milk drinking. The city—defined at that time as the island of Manhattan—provides a prime example of the development and growth of the fresh city milk industry as a whole. In the United States, New York, Philadelphia, and Boston were the earliest to experience the food supply problems of modern cities, due to rapid increases in urban population density. The older system common to smaller cities and towns—and which continued through much of the nineteenth century in these smaller centers—often involved the delivery of produce directly to consumers. Farmers delivered less perishable produce to their local merchants, who then traded with other towns and cities. Both of these systems proved inadequate for the growing northeastern cities of the mid-nineteenth century. Manhattan's island location magnified its food provision problems, and made the active pursuit of solutions to these problems particularly important.[1] As a result, Manhattan was one of the first cities to develop an organized system of urban food supply. By 1860 this system "did sustain . . . over 800,000 people on the 22 square miles of Manhattan Island, a population larger than that of 20 of the 33 states."[2] Manhattan at this time also overtook Boston and Philadelphia as the main port for transport of U.S. food exports, including cheese and butter.

By the mid-nineteenth century, "swill" milk stables attached to the numerous in-city breweries and distilleries provided the city with most of its milk. There, cows ate the brewers' grain mush that remained after distillation and fermentation. This "swill" came from the adjoining beer or liquor production facility, often poured hot into troughs directly from the brewery to the stable. As many as two thousand cows were located in one stable. According to a contemporary account, the visitor to one of these barns "will nose the dairy a mile off . . . . a high distillery, sending out its tartarean fumes, and blackened with age and smoke, casting a somber air all around."[3] Inside, he will see "numerous low, flat pens, in which more than five hundred milch cows owned by different persons are closely huddled together amid confined air and the stench of their own excrements."[4]

The brilliance of this strategy as a food waste recycling scheme was counteracted by the shortcomings of the final product: a thin, bluish fluid, ridden with bacteria, yet often sold as "country milk." While the brewery mash was in fact quite nutritious—it is still used as a feed additive today—the system was overall not a healthy one for the cow or

FIG. 2.1. Drawing of swill milk dairy barns, *Frank Leslie's Illustrated Newspaper*, May 8, 1858. From New York City Department of Health Milk Commission, *Is Loose Milk a Health Hazard?* New York: Department of Health, 1931.

its customers. The milkmen would rent a few stalls from the brewery and stable cows in these dark and dirty pens for the rest of their natural lives.

By the mid-nineteenth century, the middle classes of the northeastern cities had begun to call for the reform of this system. About the same time, some of the earliest nutritional researchers had begun to examine the value of milk. They marveled at the wide range of nutritional ingredients in this substance, and some began to refer to it as a "perfect food": containing, in perfect measure, all the ingredients to sustain life.[5]

Yet, as milk was produced in cities at the time, it was an incredibly dangerous food. Infant mortality rose precipitously in the first half of the century. By the 1840s, an infant born in the city had a 50 percent chance of living to the age of five, and the rate of death in the first year was 15 to 20 percent, leading one historian to dub nineteenth-century cities "infant abattoirs."[6]

In contrast, an infant born in the country at that time had a much higher chance of surviving to adulthood. Assessments at the time tended to point to milk and the general temptation to vice and child neglect in cities. Today, researchers also attribute higher city infant death rates at the time to milk, but in combinations with the lack of urban sanitation infrastructure, high rates of poverty, and lower breast-feeding rates, as well as basic overcrowding. Distance, one of nature's most effective forms of sanitation, was a more plentiful resource in the countryside.[7]

One of the most common urban epidemics, occurring with deadly regularity each summer, was "cholera infantum," a diarrheal disease that led to dehydration and death. Thousands of babies died of this disease every summer and, increasingly, doctors began to link these deaths to bottle feeding of cow's milk. While unsanitary water supplies must have contributed to cholera conditions in cities, milk was an even better medium. Bacteria not only survived in this substance, the protein and other nutrients provided fuel for its growth.

Experience in foundling hospitals—orphanages for abandoned babies—provided ample evidence for the problems of bottle or "artificial" feeding, also called feeding "by hand." Physicians attached to these institutions, in America and Europe, quickly noted the difference in survival rates between infants breast-fed by wet-nurses and those fed artificially. For example, of the 231 infants at New York's Randall's Island Infant Asylum, where cow's milk was the major infant food, only 26 survived. By the second half of the century, "American physicians interested in preventing morbidity and mortality among the very young were paying increasingly close attention to infant feeding."[8] Throughout the nineteenth century, physicians noted the relationship between breast-feeding and infant survival. However, without a germ theory of disease, they could not track the direct causal relationships between germs, sanitation, and the protection breast-feeding provided in preventing exposure to disease.[9]

Midcentury examinations of the infant mortality issue already expressed awareness of the fact that the rates of death were highest among the urban poor immigrant populations. Yet the problem covered all classes: New York City's burgeoning upper and middle class did not escape from the problem of cholera infantum in their own homes. The milk reformer John Mullaly's interest in the topic began with the death of his own child. In fact, in better-off New York City neighborhoods, where women were less likely to breast-feed, the effects of summer cholera were particularly acute.[10] The other contributing causes to infant mortality among the poor (bad housing and water supplies leading to poor sanitation, working or badly fed mothers) did not apply to the middle-class household, leaving milk as a more visible culprit. As a result, middle-class groups concerned with civic problems—defined primarily in terms of the reform of vice during this period—increasingly added milk reform to their list.[11]

As a result, the earliest pure milk advocate in New York City emerged from the city's temperance and mission societies. According to one historical account, the milk reform story began when an angel visited a young man who was "rapt in the ecstasies of devotion": "The campaign for a clean milk supply in New York City, which was eventually to save the lives of thousands of children yearly, may be said to have started on July 22, 1814, when a young man heard the beating of mysterious wings."[12] The young man was Robert Hartley, America's first major pure milk agitator and probably the country's first public consumer advocate. After the angel's visitation and more than a decade of prayerful reflection on the meaning of this experience, Hartley resigned his job as a factory manager in the Mohawk Valley to pursue a social reform calling in New York City. The result was a life dedicated to social reform. Alexis de Tocqueville, visiting America in 1831, described Americans as joiners of associations and as people optimistic about the prospects for human progress. Hartley filled that description well. He joined or founded many of the major city reform organizations of the time: the New York City Temperance Society, the City Mission Tract Society, and the New York Association for the Improving the Condition of the Poor. The activities of these moral reform groups would become the basis for an American obsession with the perfection of human behavior.[13]

Although an urban resident, Hartley was brought up in the upstate New York culture of the Second Great Awakening. He shared the belief of many New Yorkers that the city was a center of human degeneracy that could be bettered only through closer contact with the purer values of country people. The notion that farmers represented the bulwark of American moral life held sway through much of the early nineteenth century. It was a major component of Hartley's thought throughout his life. As secretary of the New York Association for the Improving the Condition of the Poor (NYAICP), he proposed to solve the problem of poverty by encouraging poor people to move to the country. "Escape then from the city . . . for escape is your only recourse against the terrible ills of beggary; and the further you go, the better."[14]

Yet the new future that cities represented also had taken hold of the American imagination. Hartley represented these contradictory notions. Alternately decrying the evils of modern life and celebrating the human ability for self-improvement, he reflected the contradictory thinking of antebellum America.[15]

Hartley's desire to perfect human society provided the basis for his interest in the New York City milk supply. First, as a temperance reformer, he aimed to eliminate the extra profits provided to the city's brewers through the linked milk and alcohol production systems. Second, as a social reformer interested in the uplift of the poor, he agitated on behalf of an improved milk supply. Beginning in the 1830s, in newspaper articles and lectures, and eventually summarized in a publication titled *An Historical, Scientific and Practical Essay on Milk as an Article of Human Sustenance,* Hartley exposed the unsanitary practices of the "swill milk" system. He was the first American to make a sustained argument that milk was the perfect food.[16]

Ideas about perfection emerged from the birth of American society itself, from the establishment of the American republic to the Second Great Awakening. Both revolutionary republicanism and the revivalism of the Second Great Awakening espoused the idea of the individual as a free moral agent. Republicanism propounded the idea of the moral individual against the claims of traditional monarchist authority. With the success of the American Revolution and the establishment of a democratic republic, Americans were seen, and saw themselves, as a chosen people, on a mission to create the better society for the emulation of the world. They believed that individual citizens were capable of the moral responsibility of governing a nation and therefore also the responsibility of governing their souls.[17]

Hartley was a product of this major religious revival, in which the noise of angel's wings beat up and down New York state, but particularly in the "Burned-Over District," a group of communities centered along the Mohawk Valley's Erie Canal. "Burned-over" referred not to fires but to the flames of religion, as expressed in the revival meetings that took place so frequently they became the main social sphere for much of the upstate New York populace, whether farmers, merchants, or workers, but in particular for women of the middle class.

For antebellum New Yorkers, the Erie Canal was a contradictory symbol. It stood for the ability of humanity to improve nature, creating navigable waterways to the Great Lakes by reengineering great river systems. The preachers who traveled up and down this region spoke against the old Calvinist notions of predestination. Instead, they preached the idea that people had the ability to save themselves through self-improving effort. Yet the canal also stood for what upstate Yankees would interpret as the degeneration that came with modern life's huge changes, as in the arrival of immigrant canal working men who had different social habits.[18] Antebellum evangelicals believed that Christ's return was dependent on the ridding of evil from the earth, particularly the evils of alcohol and slavery.

The rhetoric of optimism and degeneration filled Hartley's descriptions of social problems, whether he spoke against swill milk, poverty, or the evils of alcohol. This contradictory way of thinking continued in the dozens of reports he wrote as secretary for the NYAICP, which became the major charity society in the city. These discussions of the city's poor alternately described them as capable of improvement—meaning assimilation—or beyond assistance. Hartley was part of the early American welfare movement's efforts to separate the poor into "deserving" (improvable) and degenerate.[19]

Ministers preaching at the revivals of the Second Great Awakening argued against Calvinist ideas of predestination (and its corollary, predamnation), which held that an individual's moral behavior is the result of that individual's innate goodness or depravity. Instead, revivalists declared individual goodness (and vice) to be voluntary. In the words of the most prominent revivalist preacher, Charles Grandison Finney, "all men may be saved if they will."[20] Revival ministers therefore preached an evangelism of salvation to their converts, and encouraged them to try to save others. According to Thomas Pegram, "Most nineteenth century evangelicals were postmillennialists, that is, they believed that Christ would return to earth *after* the churches had stamped out immorality and defeated evil. This doctrine emphasized the perfectibility of humankind but also underlined the impossibility of tolerating wickedness."[21]

From this evangelical call came a myriad of movements for the moral reform of vice, from abolitionism, temperance societies, and antigambling societies to Magdalen societies, organized to lead prostitutes away from their immoral professions. Newly rich merchants brought up in rural communities flocked to these movements in the rapidly populating northeastern cities. Early industrialization and merchant activity at major port cities, an increase in immigration, as well as the new arrival of rural Yankees into the cities led to rapid population increases. Cities began to contain recognizable slum districts, which moral reformers targeted in their attempts to save individuals from their vice-filled lives.[22]

Reformers' experience working with the urban poor eventually convinced them that the problems of the city were not just moral but social. Hartley's career represents this move. While he founded the New York City Temperance Society in 1829, he eventually turned from this moral reform effort to found the NYAICP. Born as the social welfare arm of the Temperance Society, this organization, in its handling of poverty through self-help, assimilation, and "social visits," had a formidable influence on future American welfare policy.[23] Yet the NYAICP also represented one of the first city welfare groups in America that saw poverty as more than just a defect of character, and social reform, therefore, involving the changing of urban conditions. This change of emphasis from character to structural conditions and the move to concentration on the good but downtrodden are not accidental. As the nineteenth century matured, the Awakening emphasis on universal human perfectibility—and complete responsibility for one's own fate—became instead a job of distinguishing the perfectible—victims of circumstance—from the intrinsically degenerate.[24] This change in social reform attitude reflected an overall turn by reformers away from attention to the most "fallen" and dissipated to the reform of social conditions, in order to lead poor people away from immoral activities.

In their moral and social reform activities, groups felt the need to draw a detailed picture of moral living. While the "sinless society" was the goal, the extent of immoral activity in New York and other major northeastern cities convinced moral reformers that specific recipes for avoiding sin needed elaboration. As a result, reform societies and individual reformers published journals and pamphlets laying out prescriptions for the avoidance of sin. Not just the more obvious exhortations to avoid theaters and drinking taverns but also more specific injunctions about dress, diet, and deportment provided seekers of salvation with guidance on their every action. For example, temperance societies not only condemned liquor but also elaborated on the drinks that would be most likely to prevent intemperance, celebrating in particular the health benefits and manliness of cold water.[25]

In their search for a design for moral living, these antebellum reformers most often turned to the Bible. Even on everyday issues such as diet, reformers relied on the Bible for advice. The Bible provided the information about the natural world that reformers needed. Their ideas of social order were based on ideas of natural order, a "natural theology" that looked to the Bible as God's word about the ideal nature of social relationships. Natural theology focused on the rediscovery of God's initial design of nature at the creation. Reformers saw themselves as attempting to reinstitute God's design in the city, a place that seemed increasingly downfallen and degraded.

The publication of Hartley's *Essay on Milk* in 1842, the year he moved from temperance to social welfare reform, reflects this change in attitude. Consistent with the other moral reform writings of his time, his essay on milk paints a vision of perfection. It represents the first sustained portrayal of milk as a perfect food. Yet his essay is not just about a food; it is a treatise on the idea of perfection itself. It is not surprising, therefore, that Robert Hartley turned to the Bible when seeking evidence of milk's perfection. His biblical defense of this claim rested on two arguments: (1) consumption of milk is universal through time and space, and (2) milk is the most complete source of nutrition available. His plan for ridding New York City of its dirty milk and replacing it with pure, country milk also contained within it biblical notions of perfecting society. In a way prescient of future ideas, Hartley added two more nonbiblical components to the perfection of milk: large-scale industrialization and consumer enlightenment. The next sections will consider each of these components in turn.

UNIVERSALITY

Robert Hartley's treatise begins with a biblical exegesis establishing that milk drinking represents the perfection of human behavior, as determined by God. His interpretation seeks to show that humans everywhere drank milk. His first chapter, titled "Primary Design of Milk," associates milk drinking with the intentions of the "Prime Mover," God. Starting with Genesis, he assumes the use of this food soon after the fall from Eden, "certainly the result of design . . . man's dominion over the creatures and consequent appropriation of all they afforded was evidently the fulfillment of an original purpose of Providence in its beneficent regard for a being so frail and dependent," with the result that milk,

> for thousands of years has constituted so important and valuable a part of human sustenance. Being ready prepared by nature for food, it could at once be appropriated by the rudest savage, as well as the more cultivated. This peculiarity indeed, in an unimproved state of society, before the arts were invented, and when culinary processes were unknown, was in itself sufficient to determine his choice in favor of this form of aliment before all other kinds, which required the intervention of cookery to fit them for use.[26]

In a section of the essay entitled "The Importance of the Bovine Tribes," Hartley calls attention to the importance of herding peoples in the Bible and in biblical commentaries, starting with Abel. He quotes an early biblical commentary that painted Abel's herding as a more noble profession than Cain's tilling: "Abel brought *milk* and the first fruits of his flocks as offerings to the Creator, who was more delighted and more honored with oblations which grew naturally of their own accord, than with the inventions of a covetous man whose offerings were got by forcing the ground."[27] This celebration of animal husbandry continues with bucolic descriptions of the "wandering life" in which "every inhabitant and every family was free to pasture their flocks and herds and pitch their tents, wherever fancy might direct or Providence guide."[28] In this way, Hartley painstakingly establishes the "naturalness" of milk drinking as intrinsic to human society and given to humans by God:

> By miraculous power certain species of animals are preserved from the abyss of waters in sevenfold greater numbers, than of others. Need we inquire wherefore this special interposition and indication of Divine favor? It certainly was not on account of the animals themselves, but evidently with prospective reference to the wants of the future families of man. "Doth God take care for oxen" was the inquiry of the inspired Paul; and from his own response we learn, that this care was altogether for man's sake.[29]

Hartley claimed that the practice of milk drinking was not only universal over the time of human subsistence but universal over space as well: "The ox and his kind have followed man in all his migrations. There is scarcely a country in which they are not either indigenous or naturalized."[30] This is a surprising statement, given the fact that Hartley is speaking from a continent where domesticated cattle had been introduced only a few hundred years before.

Ten years later, another milk reformer in New York City, John Mullaly, wrote a similar treatise on the problems of the city milk supply. While Mullaly was less prone to cite scripture and more prone to cite statistics, he, like Hartley, assumes the historical universality of milk, citing a French public health publication that states, "Milk is an object of great importance to man. . . . Is it not, therefore, wonderful, that in every age this liquid should have attracted considerable attention?"[31]

Why do milk reformers feel compelled to assert that milk drinking was universal throughout time and space? Hartley cites evidence that, even if we take the Bible as an accurate account of ancient Israelite food habits, requires us to believe that the agricultural practices of Cain and Abel represented all cultures. This is not simply an ethnocentric gloss, it is a genuine attempt to establish an intrinsic natural—and God-given—relationship between being human and consuming milk.

Hartley and subsequent milk reformers spend much of their time examining, with the scientific methods of the day, the dangers contained in this food as it was produced under the swill milk system. Yet, while decrying the use of this dangerous product, they also declared milk's indispensability to society. The object of their treatises was to call upon the public to make a necessary food safe to drink.

Why do milk reformers feel the need to recommend a particular food as essential, despite its deadliness? Why is the "completeness" of milk so important? Why is a return to breast-feeding de-emphasized in their treatises?

To unravel the reasons behind this idea of milk as an intrinsically desirable and necessary food for humanity, we must first note some significant evidence against the indispensability of milk to human societies. The most striking of these is the fact that people in a large part of the world, including indigenous American, East and Southeast Asian, and many African peoples, did not milk domestic animals. These "non-milking" groups do not tend to have the gene that allows the production of lactase—the enzyme that digests lactose sugar in milk—after childhood. Contemporary studies of American populations show a vast difference in the ability to digest lactose, from a high of 80 to 90 percent in northern Europeans to a low of 40 to zero percent among Black, Asian, and Native American populations. According to one estimate, two-thirds of the world's population is lactase-deficient after the age of six.[32]

Groups that tend to have the lactase production enzyme in adulthood evolved under particular, and unusual, environmental conditions. Saharan nomads lived in an environment with little other food availability. They drank nothing but milk and blood nine months of the year. Northern Europeans in the frigid zones also had less access to a wide variety of foods. Dependence on milk to supply needed nutritional ingredients would have tended to select lactose-tolerant individuals. The prevalence of the lactase production gene in both Saharan and northern European population indicates that milk consumption is a long-standing food habit in these groups, given the amount of time it takes to select for a genetic characteristic. Out of a world of people, only these restricted groups appear to have been fresh milk drinkers for a long period of time. Fresh milk drinking has been anything but universal, either in time or in space.[33]

In the colonial era and during the early republic, Americans did not drink much fresh milk. Although most freemen were northern European by extraction, these early Americans consumed dairy products primarily in their preserved state. Milk was not an ideal beverage for New York City's hot summer days. Without refrigeration, typical July city weather could sour a pitcher of milk within hours. Therefore, while most city residents welcomed a cold, fresh glass of milk or a dish of ice cream on a hot day, such food experiences were rare, and sometimes deadly, luxuries.[34] For example, the 1850 death of President Zachary Taylor is often attributed to a glass of milk drunk after a Fourth of July celebration, after he had dedicated the cornerstone of the Washington Monument.[35]

In the countryside, milk was safer but only available part of the year. Cows generally gave milk in the spring and summer, when pasture was plentiful. During the fall and winter months, when pasture was scarce and calves were weaned, cows were "dry": they did not lactate. In winter, if people ate dairy products at all, they ate preserved dairy products such as butter and cheese. Farmers did not have extensive winter storage systems for feed or fodder. Instead, they depended on pastures to feed their animals when they were giving milk and depended on the animal's bodily reserves to keep them through the winter. As a result, "the animals became scrawny when frost withered the grasses and snow covered the earth,"[36] and those that were not slaughtered in the fall or did not die of starvation during the winter were in a severely weakened state by spring. To have one's cows "on the lift" in the spring "was considered no disgrace."[37] Under such a dairy production system, year-round daily milk drinking was simply not possible. Fresh milk drinking, when it occurred at all, was possible only during those times of year when the cow was "freshened" (giving milk because she had calved) and eating pasture grasses. Farmers also milked cows fewer months a year than they do today, generally from March to November.[38]

In fact, most of the dairy products consumed around the world are not consumed in fresh fluid form. In an overview of a series of historical studies on dairy food habits throughout Europe, the historian Patricia Lysaght concluded that in some countries "fresh milk was considered a luxury food, while in others it was thought to be unwholesome and unhealthy. In fact, it seems that the drinking of fresh milk was the least important aspect of milk utilization overall."[39] Even the "milk and honey" mentioned in the Bible is a reference to a form of fermented yogurt beverage. Fermented milk products contain much less lactose and last longer. Studies of the history of dairy product consumption in the Netherlands, Sweden, Scotland, Hungary, and Germany all show that "ordinary sweet milk was seldom drunk fresh." For example, in the Netherlands "[p]ure milk, that is, in its unpreserved form, was not an important ingredient in the medieval kitchen, because it was highly perishable, and secondly because it was not unconditionally considered healthy."[40] The food historians Gibson and Smout examine contrary arguments about the extent of milk consumption in Scotland. They conclude that the Scottish diet contained little milk until the turn of the nineteenth century, when consumption rose mostly in rural areas.[41] They cite orphanage food accounts, which contain little evidence of milk consumption, especially compared to ale. For example, in an Edinburgh orphanage in 1739, the children "were allowed a mere fifth of a pint [of milk] a week compared with six pints of ale." The Scots generally drank their milk in a soured form called *soor dook.*[42]

In most of Europe, dairying involved significant amounts of cheese and butter making, with much of the resulting buttermilk and whey fed to other livestock. Northern Europeans in America did likewise, making butter and cheese with their milk and feeding the by-products to their pigs. It was usual to keep one pig for every four cows, as a way to dispose of the by-products of butter and cheese making.[43]

Americans, like their European counterparts, most likely drank their milk in a fermented, yogurt-like, soured form often referred to as "clabbered" milk. Because both fresh and fermented milks were often just referred to as "milk," it is difficult to ascertain the amount of fermented versus fresh milk drunk during this time. The temperance reformer Benjamin Rush, one of the first to lay out detailed advice on moral living, recommended "Bonnie-clabber" milk as a substitute for alcoholic drinks.[44]

There was one group that did tend to drink fresh cow's milk more often: children. Infants could not survive on fermented buttermilk. When milk was available, older children also benefited nutritionally from the addition of milk to the diet of mainly cornmeal and pork products eaten by most Americans on the eastern seaboard in the early republic. Yet it is once again unclear whether older children at the time drank mostly "sweet" milk or the fermented form. Plantation slave children, for example, were fed primarily on cornbread added to buttermilk, probably a by-product of butter production for the slave-owner's table.[45]

Therefore, except for the African nomadic cattle-herding cultures of the Sahara, fresh milk as an essential form of daily, year-round nutrition came about only with the rise of the city. The link between fresh milk drinking and the rise of an urban, industrial society is not accidental; the daily drinking of raw, fresh milk—a food seemingly so unprocessed, primal, and "natural"—is a modern practice made possible only through the development of an industrial food system.

Perspectives on milk as an "essential" or "perfect" food seldom take this geographical specificity of daily fresh milk drinking as a food practice into account. Hartley assumed that all cultures either have traditionally drunk milk and need to continue this practice or that they should drink this God-given food. Yet even in the city where Hartley was writing, few people—besides certain infants and children—were drinking fresh milk on a regular basis. Average per person milk consumption in New York City was low, less than a third of a pint a day, including milk used in cooking and coffee.[46]

What did most Americans drink in the nineteenth century? Good spring water was "so highly thought of that a farmer bragged of a good supply, and sampled that which his neighbors gave him as an Englishman tastes tea or a connoisseur of wines sips a rare vintage from his friend's cellar." After water, the most common drink was alcohol, particularly cider:

> In the cellars of the well-to-do, a barrel of cider was always on tap, a pitcher full was on the table at every meal, the day was started with a draught of cider and was ended in the same way. The total consumption of this beverage was prodigious in homes and it was an article of export to towns, cities and even to the West Indies. Individual farmers made from 25 to 50 barrels of cider. In 1767, it was estimated that 1.14 barrels of cider per capita were drunk in Massachusetts; probably as much was used in New York.

Cider was quoted as an exchange commodity; apple juice and cider "were legal tender for the cobbler, the tailor, the lawyer, the doctor." As one commentary from the time put it, cider was a "considerable object to a farmer. The apple furnished some food for hogs, a luxury for his family in winter, and a healthy liquor for himself and his laborers all the year." From these figures and commentaries, it appears that cider, not milk, held the central place on the American table into the nineteenth century.[47]

COMPLETENESS

The milk reformer's second argument in favor of the milk-drinking habit was that milk represented a complete form of nutrition. Once again, Hartley supports his assertions with reference to natural theology. In particular, his essay quotes heavily from the British chemist William Prout. In his 1834 book, *Chemistry, Meteorology, and the Function of Digestion: Considered with Reference to Natural Theology,* Prout discusses milk's nutritive qualities using a mix of scientific and biblical authority typical of natural theology discourses of the time. Based on these researches, Prout proclaimed that milk was "the most perfect of all elementary aliments":

> Being a natural compound of albumen, oil and sugar, which constitute the three great staminal principles that are essential to the support of animal life, it is a model of what a nutritious substance ought to be,

> and the most perfect of all elementary aliments. Such being its characteristics, it possesses both animal and vegetable properties, and naturally takes its place at the head of nutrient substances.[48]

Natural theology saw as its goal the discovery of God's design in nature. This approach to the reconciliation of the precepts of religious faith with those of scientific reason is rooted in the thirteenth-century Scholastic thought of St. Thomas Aquinas and his rational arguments concerning the existence of a perfect God who maintained the order of the universe. The two religious revivals of the Great Awakening, in the early eighteenth and early nineteenth centuries—paralleled by new scientific discoveries—led to a renewed faith in the discovery of God's design as a way to determine how to live one's life.

Natural theologians, awed by the marvelous adaptation of structure to function that they were discovering in nature, concluded that there must have been a creator, or designer. They believed that the creator had made the world, and all things in it, according to a perfect, initial design: Eden. The biologist/clergyman John Ray epitomized this view. In his book *The Wisdom of God Manifested in the Works of the Creation* (1691), he stated that the objects of nature were "created by God at first, and by him conserved to this Day in the same State and Condition in which they were first made." Consequently, the role of science was to uncover God's perfect design. This attitude is also evident in Newton's declaration upon discovering the plane of rotation for the planets: "For it became who created them to set them in order. And if he did so, it's unphilosophical to seek for any other Origin of the World, or to pretend that it might arise out of a chaos by the mere laws of Nature."[49]

The argument of nature as God's design found consummate expression in William Paley's *Natural Theology* (1802). In this widely read treatise, Paley makes his "watchmaker" argument: that one can assume the hand of God behind nature as one assumes the hand of a watchmaker in the creation of a watch. It is Paley's concept of nature that forms the basis of William Prout's thoughts on milk, as cited by Hartley. In fact, Prout calls milk the ultimate proof of God's design in nature:

> Of all the evidences of design in the whole order of nature, milk affords one of the most unequivocal. No one can doubt for a moment the object for which this valuable fluid is prepared. No one can doubt that

> the apparatus by which milk is secreted has been formed especially for its secretion.[50]

He goes on to critique the more Lamarckian view that the "apparatus for the secretion of milk" may have arisen from the "wants or wishes of the animal."

> On the contrary, the rudiments of the apparatus for the secretion of milk must have actually existed in the body of the animal, ready for development, before it could have felt either wants or desires. In short, it is manifest that the apparatus and its uses, were designed and made what they are, by the great Creator of the universe; and on no other supposition, can their existence be explained.[51]

This belief in the return to Eden necessitated a static view, not only of nature, but of society itself. The search for a return to God's perfect design of nature parallels a search for God's ideas of perfect human behavior. In each case, the search was for the single, changeless, universal, perfect form. Until the advent of Darwin, natural theologians sparred continuously with the ideas of Lamarck. While wrong in his mechanism, Lamarck was radical in suggesting a dynamic mode of change in nature. In a world based on static order, Lamarck's notion of desire as the modus of change was particularly revolutionary, more so than Darwin's ideas of competition and natural selection. In the hands of the evangelical revivalists, the critique of Lamarck becomes a social prescription as well. To follow God's design does not involve following one's own desires but discovering one's duty and role in society. For Prout's investigations of digestion, this involves even the discovery of what God wants people to eat. A perfect being would give us perfect food, and milk is the food that Prout identifies with these qualities:

> After all his cooking and his art, how much soever he may be disinclined to believe it, is the sole object of his labor; and the more nearly his results approach to this object [milk], the more nearly do they approach perfection. Even in the utmost refinements of his luxury, and in his choicest delicacies, the same great principle is attended to; and his sugar and flour, his eggs and butter, in all their various forms and combinations, are nothing more or less, than disguised imitation of the great alimentary prototype *milk*, as furnished to him by nature.[52]

These early milk boosters call milk perfect because it presumably contained all the elements necessary for life. It is this completeness that makes it so important. Based on the natural theologians' claims, the completeness of milk becomes the rallying cry of milk reformers. Ten years later, John Mullaly's exposé of the swill milk trade states matter-of-factly and without any need for demonstration that "Good milk contains, as is well known, all the elements necessary not only for the nutrition, but the growth of the body." Both Hartley and Mullaly spend time discussing the chemistry of milk, the amount of albumen and "ash" it contains, and the importance of these ingredients to human health. Mullaly does not quote the Bible as an authority. Instead, his evidence is children's bodies, their mortality and, particularly, their teeth, which, due to drinking swill milk, had become "so soft that in some cases they [could] be cut with a dentist's instrument." Yet, while degraded milk was the problem, perfected cow's milk was presented as the only solution to these children's health problems.[53]

THE INDUSTRIAL VISION

Hartley was writing as much against alcohol as in favor of pure milk. In his vision of a pure milk supply, he had a clear alternative to the distillery-based dairy system, which he laid out in a chapter titled "Appeal to Farmers." He notes that, due to the increasing exposure of the problems of distillery milk,

> there is now a growing demand for pure milk produced from natural food. Such, briefly, being the attitude of this community in relation to the business, a fine opening is presented to men of enterprise who live in grass regions, within a convenient distance of the city. . . . Why should you not engage in this branch of business? Many of you are the proprietors of some of the finest grazing farms in the world, which are already stocked with cows. These you can turn to immediate and profitable account. The conversion of your milk into butter or cheese, with the loss of the labor of making it, will not pay more than two cents a quart, for which you may realize six cents in these cities. Is not this sufficient pecuniary inducement for you to engage in the business, aside from the human consideration that such an enterprise will probably be the means of saving the lives of thousands of innocent children, and of

> warding off numerous evils [the distilleries themselves] which now afflict and oppress the population?[54]

Hartley in fact visited upstate farmers, seeking to convince them of the profitability of selling milk to the city.[55] Yet the building of railroads into New York City lagged behind that of many other cities at that time, making New York City's overall food supply higher-priced and of poorer quality than the food available in many other cities. Spann describes the relationship between the development of the rail system into New York City and the consequent improvement in food supplies:

> Before 1845, supply by land was limited to farm wagons, droves of cattle herded on the hoof, and New York's still feeble rail system. Railroad development after 1845 rapidly opened up new inland areas. By 1850, besides its connections with New England, New York was served by some 762 miles of railroad; by 1860 the figure was 1110. The railroad may not have increased the area of the city's food supply by the 4600 percent estimated by rail enthusiast Henry Varnum Poor, but the new system did effectively supplement, and increasingly supplant, the older water and land routes by providing faster and more flexible service.[56]

Ten years after Hartley's treatise, one third of the city's milk supply was brought in by rail. This new type of dairy industry was heralded by John Mullaly and the next generation of milk reformers as the alternative that would achieve the goal that Hartley originally set out: to reunite the city public with its biblical fresh milk drinking legacy. This new system brought "country milk" over the newly constructed rail lines into the city. As subsequent chapters will explore, the link between the development of railway infrastructure and the provision of this "perfect food" to the public is just the first in a series of links made between industry, technology, science, and milk drinking.

ENLIGHTENMENT

A thread that can be followed all the way from Hartley's original treatise to the National Dairy Council's education programs today is the idea that consumers need to be enlightened about the perfection of milk and about its dangers. The early reformers told many stories about how milkmen who sold swill milk from city distillery dairy barns assured their customers that their product was "country milk" from upstate Orange County. For example, Hartley noticed,

> Some men in the milk business, very reluctantly forego the profits of slop-feeding, and rather than lose their customers, resort to various subterfuges and evasions. . . . A gentleman says his milk man assured him that *he fed no slop*; and as no evidence appeared to the contrary, he felt bound to believe him. But passing a distillery some months afterwards, curiosity induced him to stop, and, to his surprise, he saw his milkman busy among his cows. As the truth flashed across his mind, he charged the milkman with deceiving him, who promptly replied: "Every word I said was true, sir; I told you *I fed* no slop; and by the help of the gutter, you see, which leads from the still-house to the stables, my *cows feed themselves!*"[57]

Caricatures of this deception pictured the "country milk" wagon filling up in front of the distillery dairy.

Reformers fought the swill milk system through public education campaigns. Their efforts, however, were met with "a degree of apathy abroad on the subject of milk-food, from which it has hitherto been impossible fully to arouse the public mind" against the continuing sale of swill milk in New York City.[58] Consumer education was an intrinsic part of the reform equation. It was also part of the idea of perfection: once the perfect way was discovered, the only human intervention necessary was to direct the multitude toward this goal.

The enlightenment story combines the perfection of industry with the perfection of consumer knowledge. For example, Spencer and Blanford attribute the increase in milk drinking to "significant improvements in the quality of milk and cream sold," which led to a "more generous use of those products." Consumers drank more milk because they had "greater knowledge of the food value of milk,"[59] which was the result of "favorable teaching and publicity based on important findings and research."[60] In other words, the rise of milk consumption, according to these economic studies, is due to the increasing perfection of milk—in both quality and price—and education of consumers about this perfection. The history of milk drinking becomes a history of this increased perfection through increased consumption and through a public/private promotion of the product.

Spencer and Blanford's explanation for the rise of milk drinking is not really an explanation at all; it is a reference to a particular idea about perfection and how, over time, society has marched in the direction of this goal: more milk drinking. The idea that American milk drinking may be a product of social choices is not allowed into the terms of discussion, from the viewpoint of milk's perfection. After all, if milk is perfect, then the only role society has is to drink more of it, and to make sure others drink more.

Questioning the perfect milk story opens up new questions, and explanations for milk drinking no longer appear as autonomous forces. Instead, social interaction, vision, and political action come to the fore in a more "social" view of society. The explanations that rely on "important findings of research" and the "favorable teaching and publicity" become open to scrutiny. The more open, social story enables us to take a closer look at the way milk's actors created this food. In particular, the relationship between what people eat and what they "know" becomes clearer. The idea of milk as a perfect food was born in a particular way of knowing the world, and a particular way of understanding the divine.

THE DREAM FULFILLED

These mid-nineteenth-century milk reformers were describing a dream: a city in wait for fresh, pure milk. The reformers' dream was to reunite a populace with its "natural" diet, designed by God. Yet, while they placed this vision in a biblical past, they were in fact envisioning the future. In this future, consumption of this "perfect," "complete" food would become a major American food habit. Because this vision in fact came to pass, it is more difficult to imagine their vision in the realities of that time: few people drank milk, children drank most of it, and most of it was not safely drinkable.

This dream of a milk-drinking society did become a reality. By the 1950s milk had in fact become one of the most widely consumed foods in the United States. Even today, Americans on average consume over five hundred pounds of dairy products a year, comprising the largest group, by weight, in the nation's diet.[61] Yet, while the perfection of milk has come under question recently, for many decades the idea that contemporary milk consumption habits reflect universal laws about the perfect human diet went unquestioned. Surprisingly, the rationale behind the perfection of milk—despite decades of evidence to the contrary—still revolve around the two major arguments propounded by Hartley in antebellum New York City: universality and completeness.

A typical example is a 1950s book by the dairy economists Chester Roadhouse and James Henderson, *The Market-Milk Industry,* written more than a century after Hartley's essay. Written for a milk industry audience, the authors state that they are laying out the facts about milk nutrition so that the industry person can "answer the questions of customers" and appreciate "the importance of his product to the health of his community."[62] While most of the book is about the economics of the industry, the authors include an eight-page history of the industry and a discussion of milk as a food.

Their chapter begins with *the* typical booster statement: "The milk of animals was used for human beings before the dawn of history." It goes on to call milk "the most nearly perfect food." Ever since Hartley, these constantly repeated phrases echo down through the American literature on the nutritive value of milk. The "special value" of this food comes from the fact that milk's role is to "nourish the species before a mixed diet can be taken" and for this reason it "occupies a unique position." It is, therefore, "the one food that supplies most of the nutritional requirements in the proper form and balanced proportions."[63]

Needless to say, Roadhouse and Henderson do not cite the Bible as evidence for the universal consumption of milk. By the 1950s, people talked about vitamins and minerals to describe the chemistry of milk. Darwin had replaced God's design as an explanation for the natural world. Governments closely regulated milk production and price. However, the modern argument for milk drinking is strikingly close to those stated before the discovery of vitamins, germs, Darwin, or dairy regulation: milk is the first food to sustain the human body, and it contains within itself the universe of nutritional needs. Therefore, milk is not only universally used over time and space, but contains within itself the universe of nutrition. These three universals: history, geography, and the commodity itself, make milk boosterism almost a spiritual quest for the unification of time, space, and the body. While Hartley's claims may seem more outlandish and less well informed, his final goal is not so different from that of Roadhouse and Henderson. Hartley is attempting to reunite a city population with its milk-drinking history and God's divine will, while Roadhouse and Henderson are instructing a milk distributor on how to educate the public on "the importance of his product to the health of his community." In both cases, the goal is to induce in the public the habit of milk drinking, for its own good.

Roadhouse and Henderson praise the milk industry and government for providing this product safely to the public. Their milk history, a true progress narrative, lists the people and institutions that contributed to the creation of the milk provision system. Their introductory chapter on the history of the market milk industry includes a section titled "Educational Institutions Contribute to Dairy Progress," which describes the role of the land grant institutions, established by an act of Congress in 1867 "to initiate research and disseminate knowledge concerning agriculture and dairying." The next section of this chapter, titled "Science Laid the Foundation of Progress," describes how the sciences of bacteriology and chemistry "emphasized the importance of sanitation and sterilization of dairy utensils and facilitated the rapid progress made in the market-milk industry during the past thirty years." These discoveries included Leeuwenhoek's first observation of bacteria through his improved microscope in 1683, the discovery of lactic fermentation in 1857 by Pasteur, and his disproving of the theory of spontaneous generation, which led to Koch's establishment of the relationship between bacteria and disease. Koch's work was specifically related to tuberculosis and his effort to develop a method for diagnosing tuberculosis in cattle.[64]

These economists go on to describe the "Conditions Leading to Improved Quality," including the efforts of milk reformers (the next generation of Hartleys and Mullalys) who lobbied for the pasteurization of milk at the turn of the century, which I will describe in chapter 4. They go on to describe other developments that made the provision of fresh milk to the cities possible. This included the development of the glass bottle by Thatcher in 1886 ("the new bottle provided a means for protecting milk from contamination after processing") and the development of medical milk commissions, whose standards for "certified" milk "set a high standard for sanitary production and quality of all milk during the years that followed." They also note the development of "tank cars and tank trucks for transporting milk to the cities" as well as the invention of mechanical refrigeration, made possible by Jacob

Perkins's discovery of the compression cycle in 1834: "No invention has been more important to the dairy industry."

Finally, they cite the establishment of the "health supervision of milk supplies in the United States," which originated with the first milk ordinance, passed by Massachusetts in 1856. New York passed its own ordinance for the inspection of milk in 1866 to prevent adulteration, and the city banned swill dairies in 1873. In 1896 the Board of Health of New York adopted a permit requirement for all sale of milk in the city, which Roadhouse and Henderson call "the beginning of the present effective system of controlling and protecting the milk supplies of cities."[65]

At the turn of the century New York City saw "rapid progress in milk supervision." In 1906 Dr. Thomas Darlington, the city's health commissioner, instituted the city inspection of dairy farms, partly in response to decades of lobbying by Hartley's social welfare organization, the NYAICP. He began the system of dairy farm and plant inspection, which became the norm in most cities by World War I. By the time of Roadhouse and Henderson's writing, the industry had achieved "a high point in sanitary production, transportation, processing, and distribution." In addition, "consumption of milk per capita has markedly increased during recent years." More than one-third of all milk produced went into the market milk system by this time, compared to only a few percent in Hartley's time. By the time Roadhouse and Henderson are writing their book, the fluid milk industry had become "the most important branch of the dairy industry in value of products, investment in buildings and equipment, and the number of people required to do the work."[66]

Roadhouse and Henderson's eight-page history of milk summarizes many boosters' themes—and many of the themes of this book—not only for milk, but for modern food in general. It is a modern story of progress that is never really without foundation. Technology, science, and government intervention did give us the ability to drink fresh milk every day. The question is, why? The boosters' answer to that question is a declaration of the "perfection" of milk. Declaring a food perfect is tantamount to declaring it necessary, and if it is necessary, the system that brings it to us must be necessary as well. Yet it is that rhetoric of perfection that has kept us from taking a more questioning look at the history of our milk system and how it has been structured.

Practically any history of milk follows the structure of this progress narrative. Shaftel, for example, begins his history of milk regulation in

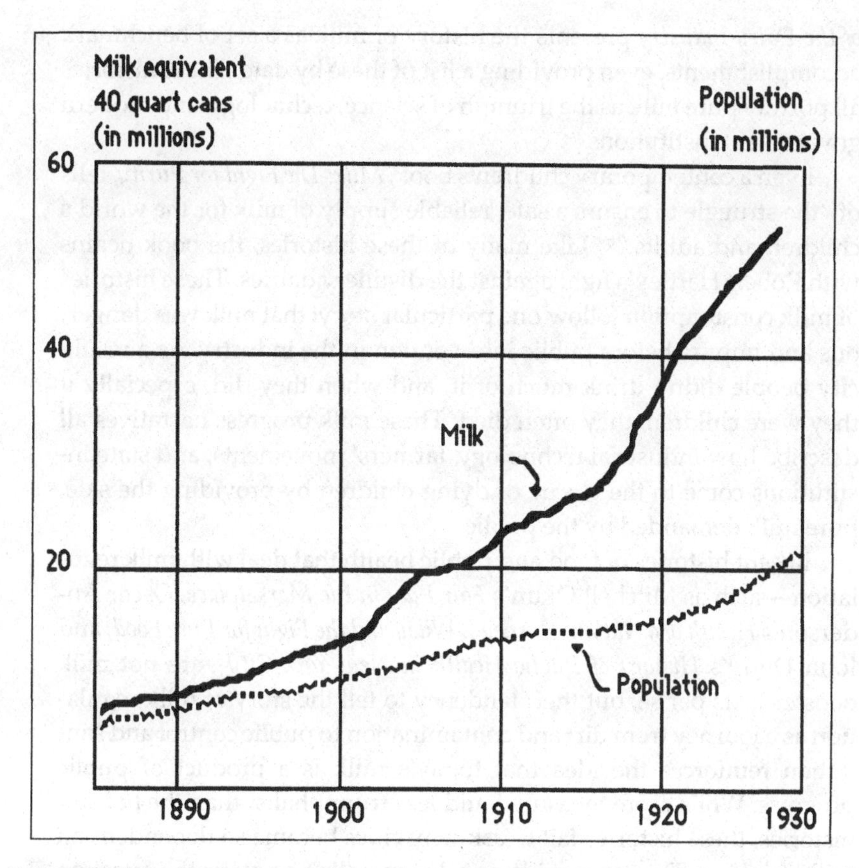

FIG. 2.2. Approximate population of the New York City area and receipts of milk and cream at the New York market, 1885 to 1929. From Catherwood, *A Statistical Study of Milk Production for the New York Market.*

New York City by stating that the story is one of "righteousness born of knowledge, triumphant in the end," in which "the developments and advances in milk purification will be discovered to follow in causal relationship social, political, and scientific maturity."[67] For example, Catherwood's *A Statistical Study of Milk Production for the New York Market* displays charts showing the increase in the amount of milk delivered to New York City per capita between 1885 and 1929, a calculation commonly used to define historical trends in milk consumption in a city (fig. 2.2).[68] Dairy boosters portray this rapid rise of milk consumption as a product of pasteurization and consumer education. T. R. Pirtle's *History of the Dairy Industry* presents the history of milk as a set of benchmark accomplishments, even providing a list of these by date. These histories all portray pure milk as the triumph of science, technology, and modern government institutions.

Even a contemporary children's book, *Milk: The Fight for Purity,* tells of "the struggle to ensure a safe, reliable supply of milk for the world's children and adults."[69] Like many of these histories, the book begins with Robert Hartley's fight against the distillery dairies. These histories of milk consumption follow one particular story: that milk was dangerous and impure before public intervention in the industry. As a result, city people didn't drink much of it, and when they did, especially if they were children, they often died. These milk progress narratives all describe how industrial technology, farmers' movements, and state institutions come to the rescue of dying children by providing the safe, pure milk demanded by the public.

Recent histories of food and public health that deal with milk regulation—such as Mitchell Okun's *Fair Play in the Marketplace,* Oscar Anderson's *Health of a Nation: Harvey J. Wiley and the Fight for Pure Food,* and John Duffy's *History of Public Health in New York City*—are not milk booster texts per se, but their tendency to tell the story of milk regulation as a journey from dirt and contamination to public control and sanitation reinforces the idea that today's milk is a product of public progress. While more measured and less triumphalist than the booster histories, these histories fail to ask why cities became so dependent on milk drinking. This type of historical storytelling answers the question "How did we get to where we are today?" without ever asking, "Why are we here?" While these stories give invaluable information about the state of present food habits, their dependence on the current situation as the end goal of the narrative avoids any examination of alternative possibilities.

CONCLUSION

If we disentangle the story of milk from its progress narrative, it is possible to tell a more social story about milk, one that seems less inevitable and makes the consumption of this food a product of choices made rather than an ideal fulfilled. As this chapter shows, social ideas and human action were very much "present at the creation" of milk drinking as an American habit. The human actors at the time were not capitalists manipulating public sentiment and choice for their own purposes, although these forces will become increasingly present in the rest of the story. The system began with a vision of perfection and a critique of the system that then existed to feed infants. The vision was free of personal interests or profit, but it was not free of power and ideology. Reformers had strong ideas about the form a "good" society should take. That vision led to specific ideas concerning the relationship between food and the body. Yet the source of that ideological hegemony was temperance and piety, very different from the later ideological manipulations created by industry advertising and government milk promotion.

There are many possible ways to explain the role of this religious vision in the beginnings of the industrial milk system. One possibility is a spin-off of the Protestant ethic: just as Weber found the spirit of capitalist enterprise in the religious beliefs of Calvinism, the spirit of capitalist consumption practices may have preceded their fulfillment through the rise of the industrial food system. The "spirit" of consumption was present before the rise of production to satisfy this spirit, a spirit that expressed itself as a social movement, that is, as a form of politics, with other countervailing politics.[70]

The progress narrative of milk tends to ignore the social parts of this food's history. Choices made, social contexts within which particular circumstances occurred, and events that do not fit the story are either ignored or explained away. In other words, this form of storytelling ignores particular groups of people by excluding them from the story or by explaining their activities as a less perfect form of life. Only once we have freed ourselves from the progress and downfall narratives can we begin to see, and have a way to explain and understand, the role of ideas and power in the history of milk.

The storytelling in the progress narrative tends to follow a particular structure, which tends to close off the question "Why milk?" Retelling the story of milk therefore requires a careful examination and questioning of this narrative. To do this, we need to question the claims made by milk reformers about milk as a universal and complete food that large-scale industry can bring to an enlightened public. To accomplish this task, the following chapters will pay attention to the way the perfect story about milk gets told. The discussion will return repeatedly to the major strategies used to tell the perfect story:

Representation

In both words and pictures, the perfect milk story has revolved around the representation of particular ways of life as perfect, and the depiction of alternate ways of life as degenerate. In these images, some actors involved in the economic and social activity around milk are invisible while others are prominently featured. This book will focus a great deal on the common images that accompany milk history and milk planning.

Hagiography

Stories represent the people involved in the creation of milk as champions of progress. These people are described in reverent biographies—often accompanied by portraits—similar to hagiographies, or "lives of the saints" in religious literature. The story of milk is therefore told as a list of these people and their innovations.

Benchmarks

Physical indicators, particularly technologies, are commonly designated as signposts or "benchmarks" on the path to progress. Lists of the invention and application of particular technologies—Gail Borden's vacuum pan system, the milk machine, the milk tester, the cream separator—often accompanied by images, tell the story of the increasing perfection of milk. A description of dairy technology by the agricultural historian T. R. Pirtle is typical: "There has been a great increase in the consumption of milk in recent years," mostly in cities, he stated. "This increase in cities has been relatively continuous since the improvement in transportation, the invention of tinware, glass milk bottles and the application of sanitary methods, that is, the improvement of quality precedes increased consumption."[71]

Thinking by Graph

Finally, the progress story rejects social explanations of change in favor of impersonal forces that are often represented as lines on a graph. This "thinking by graph" march toward progress and perfection becomes a predefined voyage that does not require human drivers.

Exceptions are ignored or else represented as less perfect: they are either placed earlier on the progress timeline (when they often are coterminous events), or treated as temporary retrogressions, due to evil intentions or ignorance, that held back history for a time.

The historical line graph follows the rise or fall of a particular historical phenomenon. This type of graph shows the growth of a particular activity, which is accompanied by an explanation that sees the activities at each stage of the graph as an incremental movement toward the perfect goal. In the story of milk, the goal was the provision of a safe version of this food to American cities. Historians illustrated the story with graphs such as the one in fig. 2.2.

In their book on the industrialization of food, *Refashioning Nature,* David Goodman and Michael Redclift argue that, to industrialize food, producers needed to physically change the food itself. Thinking by graph mystifies this point. The chart hides the fact that the milk commodity represented at the end of the chart is an entirely *different* food from that represented at the beginning of the chart. Thinking by graph is a form of perfectionist thinking that this book will attempt to dismantle, in order to tell another story about milk.

3

Why Not Mother?

The Rise of Cow's Milk as Infant Food in Nineteenth-Century America

GIVEN THE UNHEALTHINESS of the product in the mid-1800s, why did anyone at all, much less vulnerable infants and children, drink cow's milk? Accounts that celebrate the perfection of "country" milk and the poisonous nature of swill milk tend to ignore another widely available and widely used source of pure milk at that time: the human mother. Hartley's hundreds of pages linking humanity to milk drinking and animal herding ignore the basic historical fact that the milk drunk most commonly by humans across history and geography has been, in fact, breast milk. Mullaly's book on the milk trade extols the purity of country milk but never mentions this alternative and traditional source of milk for children. Hartley, extolling the benefits of cow's milk, describes it as "the natural food of the infant. . . . when pure it is at once the most palatable, healthy, and nutritive aliment with which our nurseries and tables can be supplied."[1] He does not mention that human women had been, until that time, supplying milk, but instead concludes that this natural food must be "supplied."[2]

In fact, authors writing in the mid-1800s can assume that milk is a complete food only because of the universal consumption of human milk from the dawn of humanity. The science of the day, however, could distinguish little difference between cow's milk and human milk. Yet the rise of urbanization and the rise in artificial feeding of children went hand in hand. The increased substitution of cow's milk for human milk led to a proliferation of reports that compare the nutritional contents of these two substances. Prout, Hartley, and Mullaly, in their writings on milk, all report on the findings of the cow/human milk comparisons, which find that cow's milk is an adequate substitute for human milk.[3]

Artificial feeding, as a practice, is not new: archaeologists have found occasional milk bottles and "pap boats"—containers for a pablum of flour and milk or water—in tombs of the Roman classical period. Instances of particular regions feeding their children "by hand" was not unheard of before the rise of cities. In certain dairy-intensive regions of central Europe and Scandinavia, feeding infants cow's milk was the cultural norm. For instance, in Norway and northern Sweden in 1825, most infants were fed cow's milk, and infant death rates there were high.[4]

Therefore, while there is a long history of artificial feeding, this long history also shows that this form of feeding generally meant death for an infant.[5] Researchers today still find some relationship between child health and breast-feeding, even in the industrialized countries. However, the health impacts are minimal compared to the effect that artificial feeding has had on a child's chances of survival in places and times where there was no knowledge of or ability to maintain the sanitary conditions necessary for safe artificial feeding. Parents in nineteenth-century cities did not have the knowledge or ability to provide the clean environment necessary for safer artificial feeding. Consequently, the rise in bottle-feeding in nineteenth-century cities was accompanied by an increase in infant mortality rates.[6] As one public health official put it, vital statistics showed "the malign effects of hand feeding of infants and the consequent impress made upon the mortality returns."[7]

Throughout the history and prehistory of the human species, breast milk provided the major sustenance for a person's first year of life. European women provided breast milk not only to infants but to invalids. It was also widely used in women's healing remedies.[8] In other words, women's breast milk production represented a significant part of the human food economy. It is an unrecognized part of women's contribution to the human food system. When one looks at women's bodily production as part of the food system, the transition from breast-feeding to artificial feeding takes on a greater significance. One does not have to be a member of La Leche League to understand that a significant economic change took place when women's bodies were removed from food production.

In the germ-filled city environments that existed before the building of city water and sewer systems, breast-feeding also provided significant child health protection, although this was mostly unrecognized at the time. Crowding in cities increased the exposure of children to disease. The interaction between crowding and artificial feeding had a deadly effect: studies show significant statistical differences in nineteenth-century rural and urban infant mortality rates.[9] Recent comparative historical studies of breast-feeding practices in nineteenth-century cities show that cities where more women breast-fed their infants had significantly lower infant mortality.[10] Even during this period, experience in nineteenth-century foundling hospitals made the relationship between artificial feeding and infant mortality rates strikingly clear.[11]

Yet milk reform texts betray multiple suspicions about the healthfulness of women's breast-feeding. Robert Hartley's treatise, over 350 pages long, mentions breast-feeding only twice. In the first section, titled "Popular Mistakes," he follows a lengthy discussion on the milk of diseased animals with a query about the health condition of breast-feeding mothers: "Who does not know, for illustration, that the health of the infant is affected by the condition of the sustenance it receives from its mother? . . . Is the mother diseased? The virus generated in the vitiated secretions, taints the nourishment, and is communicated to the child."[12] One might attribute these statements to ignorance about the communication of antibodies and other immunities—as opposed to the potential but much less crucial transmission of disease—through breast milk. However, Hartley's statement is more striking considering that milk had been, up to the nineteenth century, used as a medicine.

But beyond the suspicion of diseased mothers—which was an exceptional circumstance—he expanded his suspicions of the quality of mothers' milk for a reason much more common and much more likely to resound with average middle-class urban women of that time: nerves. While listing diet and health as two factors that affect the quality of a woman's breast milk, he adds that "irregularities" can be attributed "probably more frequently to the influence of the mental emotions, which as they happen to be unfavorably affected, produce corresponding changes in the milk that seriously injure the health of the infant, and in some instances have proved fatal."[13] To prove this claim, he quotes an anecdote, quoted in 1836 by a doctor quoting, in turn, another doctor who was physician to the king of Saxony. In this story, a soldier who was billeted to the house of a carpenter began to argue with the man and drew his sword on him. The carpenter's wife came between the two men and wrestled the sword away from the soldier, with neighbors assisting in separating the two men.

> While in this state of strong excitement, the mother took up her child from the cradle, where it lay playing and in the most perfect health, never having had a moment's illness; she gave it the breast, and in so doing sealed its fate. In a few minutes the infant left off, became restless, panted, and *sank dead on its mother's bosom.* (italics in original)[14]

While women did not commonly wrestle swords from soldiers, they were commonly presented, and presented themselves as, "excitable" during the Victorian era, and worries about how emotions would affect breast-feeding would certainly have been linked to the then-current fashion in female temperament.[15]

While for the most part ignoring the traditional role of women in supplying milk for children, Hartley does distinguish between urban and rural women in attempting to explain the significant decline in breast-feeding in cities as they grew larger. He blames city life itself, both in its unhealthiness and in its temptations to lead women into unhealthy lives: "Many mothers in our large towns, from constitutional feebleness, and others from infirm health, are incapable of nursing their offspring; but far more from unnatural and justly reprehensible habits of life, completely disqualify themselves for discharging this important and endearing duty."[16] Once again, the appearance of sickliness was a fashion among upper-class and even some middle-class women at the time. The vague description of "reprehensible habits," however, is unclear. It is particularly disturbing considering his estimate of the number of children fed "by hand" in the city at that time: "in the judgement of several distinguished medical practitioners, whose ample opportunities for observation entitle their opinions to respect, more than *three-fourths* of the infants born in our cities are sustained in whole or in part on artificial diet."[17] Unless one imagined that nearly three-quarters of all women with children in the city at the time were either unhealthy or easily tempted to vice, Hartley's explanation is inadequate.

Hartley bases his call for milk trade reform on the harm impure milk was having on children. Yet one has to ask why the author never considered a reversion to the previous form of infant nourishment: breast-feeding. Answering that question requires looking at the complexity of women's roles in mid-nineteenth-century urban American society.

Hartley claims that fully three-quarters of all children in New York were fed "by hand," that is, with cow's milk rather than breast-feeding.

This may have been an overstatement. Recent historical research has shown that colonial American mothers generally breast-fed children for the first year.[18] While few studies of American breast-feeding rates for the mid-nineteenth century are available, late-nineteenth- and early-twentieth-century figures are available for some European and American cities. Fildes notes that in early-twentieth-century London, only 7.6 percent of all infants were artificially fed from birth, although rates of breast-feeding in both American and British mill towns were significantly lower, since many women worked in the mills.[19] In Berlin, 79 percent of all women in 1885 breast-fed for over eight months.[20] One of the first major scientific studies of infant mortality in the United States was published by Robert Morse Woodbury in 1926. Covering eight cities, the data showed that only 25 percent of women did not breast-feed their babies.[21] It is hard to imagine that New York figures in the 1840s would diverge so strongly. However, as early as 1890, the prominent pediatrician Luther Emmett Holt stated that "at least three children out of every four born into the houses of the well-to-do classes" were not breast-fed.[22] Hartley's figures, based on physicians' accounts, probably represent the mostly upper- and middle-class people who comprised the main medical clientele at the time.

However, even if Hartley's estimates were extreme, there was clearly a move toward more artificial feeding in New York City by the mid-nineteenth century. Even if New York followed the trends in other cities of the time, a larger number of children were being fed cow's milk than was the case fifty years earlier.

To unravel the reason for the breast-feeding decline in the mid-nineteenth century, one can start by examining the myriad of explanations that have been offered. One explanation *not* offered by historians for the decline in breast-feeding during this period is a version of today's infant formula controversy. Nestlé's encouragement of formula feeding for infants in the Third World has sparked great controversy in the last few decades, leading to a worldwide boycott of the company's products. Did infant formula companies, through advertising, create this same need in the 1840s and 1850s in New York City?

In fact, infant formula advertising could not be the reason for the decline in breast-feeding in the 1840s, because infant formula did not then exist. Infant formula was not on the U.S. market until the 1860s.[23] By the 1880s, high-pressure advertising, by both formula makers and the condensed milk companies, tried to convince mothers of the benefits of bottle-feeding.[24] However, these forms of processed cow's milk could not have been responsible for the initial decline in breast-feeding rates, since they did not exist at the time of Hartley's essay. Hartley called for the reform of the fresh milk provision system because babies were drinking primarily fresh cow's milk at the time.

Were local fresh milk providers, then, using advertising that would pressure mothers to buy milk rather than breast-feed? This appears also not to be the case, for two reasons. First, a perusal of newspapers and magazines published in the 1840s and 1850s shows that the local dairy industry did not advertise in these publications, in part because that sort of advertising did not exist. Later promotions of the new infant formulas represented some of the first advertisements in women's magazines. Many of these formulas, such as Eskay's Food, were created by patent medicine companies and a great proportion of the early advertisements in these magazines were for patent medicines.[25] Yet this type of advertising, like infant formula itself, only came into its own in the 1880s.[26] Second, the structure of the infant formula industry—industrial production for a "multinational" market—was vastly different from the small-scale milkman who rented a stall in a distillery barn and peddled milk on the streets. Only mass-production firms could utilize mass advertising for their products, and the growth of mass industry was reliant on the growth of mass advertising. Milk peddling did not qualify. Milkmen relied on the signs on their wagon and their milk receipts. Larger dairies advertised in the city directories, but they do not mention the use of milk as a substitute for breast-feeding.[27]

Another possible influence was the growing child-rearing advice manual industry. Were advice manuals and local experts advising women to feed their children cow's milk? No. Child-rearing advice manuals had existed as early as John Locke's seventeenth-century essay *Some Thoughts Concerning Education* and Jean-Jacques Rousseau's eighteenth-century work *Emile*. As Grant states, "[o]nly a sprinkling of child-rearing advice had made its way to America during colonial times, but by the early 1800s the supply was growing."[28] However, the vast majority of these books treated breast-feeding as a woman's sacred duty and chastised women for not performing this role. For example, Buchan's popular *Advice to Mothers* (1804) "advocated that mothers breast-feed their own babies, and disparaged the use of medicines for children."[29] In some countries, the voice of authority in favor of breast-feeding became law: the Prussian code of 1794 required that all healthy women breast-feed their children and assigned the father responsibility for determining the time of weaning.[30] Yet it appears that breast-feeding declined among the upper and middle classes despite the advice of the child-rearing authorities of the period.

There is also some evidence that upper- and middle-class men strongly influenced their wives not to breast-feed, because of cultural mores proscribing sex during the period of nursing, and because of a desire to maintain their wives' fertility. Earlier, these sentiments led husbands to hire wet nurses for their children, although artificial feeding was also sometimes practiced. By the mid-nineteenth century, the alternative to maternal breast-feeding was increasingly artificial feeding.[31]

While both wet nursing and artificial feeding, unlike breast-feeding, required monetary investment, in nineteenth-century America artificial feeding was significantly cheaper. Information on wet nurse wages is scarce, but twenty-five dollars a month was not uncommon, and did not include food, board, and clothing.[32] In contrast, milk in the midcentury cost about seven cents a quart.[33] There must have been some temptation for middle-class women to imitate the nursing-free lifestyle of their betters by substituting affordable cow's milk for the unaffordable wet nurse. In fact, remaining at home to breast-feed would have prevented a woman from maintaining the social and civic activities crucial to the attainment and maintenance of social status in urban society at the time.

For the working-class mother, however, whether or not to breast-feed was generally not a choice. Milk, whether provided by a wet nurse, formula, or cow's milk, was not affordable, and the choice to breast-feed among the working classes was often heavily influenced by a woman's economic situation: Did she fall into the category of the "desperately poor"? Was she the single wage-earner? Did she live in a manufacturing town in which the mills commonly employed women? Throughout the nineteenth century, the vast majority of working-class mothers breast-fed their children, yet under specific economic circumstances, these women resorted to artificial feeding.[34]

Among the very poor, the economics of feeding the family was influenced by a number of factors that sometimes led to early weaning. Studies in Britain have shown that women often ate substantially fewer calories than men and working children, and they often sacrificed their own food for others in the family. For example, Ross describes the situation in nineteenth-century London: "Food was scarce. Even breast milk was not very plentiful, and a woman with a "good breast of milk" was often pointed to with praise or envy. One whose breast milk failed knew that her child's life had become more insecure."[35] While nutrition was significantly better for all classes in America than in Europe,[36] the rise of destitute female heads of households must have led to some artificial feeding among the very poor. The infants of the very poor would in many cases lose access to the mother's breast through her death or abandonment. In addition, single mothers often had to work outside the house (sometimes as wet nurses), requiring arrangements for others to feed their infants.[37]

Very poor women and women of the upper classes may have had one thing in common: inadequate food intake for breast-feeding. While very poor women may have been calorie-poor, for upper-class women, dainty eating was part of the Romantic aesthetic that was pervasive by the late nineteenth century. It is not clear to what extent this was true during the 1840s and 1850s, although the fashion has been linked to the Romantic poets of the period (particularly Lord Byron).[38] Anorexia, therefore, may have had earlier roots than the Victorian era. Thus it is possible that the cause often cited for feeding cow's milk at the time, namely, insufficient milk, may have in fact been the case for women in two diverse classes.

For working-class women, another possible explanation for the decline in breast-feeding is increased participation in the workforce.[39] Women in mill towns tended to wean very early, because they went back to work very soon after childbirth.[40] In other places, married working-class women usually worked at home (often employing younger children in their tasks). Women who worked as domestic servants or factory workers were mostly unmarried or without children.[41]

None of these explanations capture the complexity that must have been involved in the decline in breast-feeding in mid-nineteenth-century New York City. The lack of explanatory power of each story reveals the problems with each causal perspective. Explanations that emphasize the power of the father to influence breast-feeding practices tend to come from authors who center their historical viewpoints on the patriarchal structure of Western society. Explanations that emphasize employment pressures on working-class women are based on a more structural explanation of history. Explanations that center on upper-class women's role in society as a reason for the decline in breast-feeding derive from more Weberian notions of status and class. Each of these perspectives, and the explanations that derive from each point of view, most likely contain some truth. However, we need to weave these explanations together to arrive at a perspective that focuses on broader societal changes and changes in women's identity during this period.

It is impossible to consider the decline in breast-feeding in mid-nineteenth-century New York City without looking specifically at the extraordinary changes women—and men and children—faced during these decades. In particular, the rural family of the early 1800s worked together, often near other family members or within larger community networks. However, by midcentury, New England families were becoming increasingly mobile, both geographically and socially. Both forms of mobility generally took these families to the city. For middle-class men, moving to the city generally meant engaging in a larger trade or professional business. But for middle-class women, the move to the city meant the abandonment of a women's economy for the private sphere of the home. Women's rural production economies—including the women's breast-feeding economy—were based on long-standing local relationships. Yet in the city, "the stranger became not the exception but the rule,"[42] women's economies broke down, and breast-feeding became an individual, not a social, pursuit. The headlong rush into the cities that began early in the century left women of every social class in an extraordinarily isolated economic and social position.

Rural economies in the colonial era and during the early republic depended heavily on work "swapping," particularly in the case of women's production. Women during this period were enmeshed in local "helping out" networks. They swapped household, child care, and agricultural production tasks.[43] These networks made little distinction between private and public economic spheres: work swapping included all the aspects of family reproduction—from birth to death—as well as more commercial endeavors such as cheesemaking. For example, women formed cheese "circles" in which one woman took in the milk of several farms and made the cheese for that day.[44] They also cared for one another during and after childbirth. These networks of obligation included informal, temporary breast-feeding of others' children. A common myth made this breast-swapping necessary: a mother's early milk, colostrum, was considered unhealthy for children. Therefore, newborns were commonly fed by others until mothers' "real" milk came in.[45] Therefore, American mothers in the colonial era and during the early republic had access to a more informal "wet nurse" network through their participation in the local women's economy. "These services were probably exchanged regularly, a part of the mutual support women gave and received during their childbearing years."[46] As one account at the end of the eighteenth century described it, "Either in town or country, . . . breast milk may be obtained from some relative, friend or other healthy woman . . . as the sex of all ranks feel much for each other in this situation."[47]

While this mistaken view of the unhealthiness of colostrum must have placed newborns at greater risk, the immediate initiation of an informal breast-feeding network around a mother and her child must have also been helpful in later circumstances when, due to various reasons, women were unable to breast-feed.

With the rise of the urban middle and upper classes, wet nursing moved from the realm of informal mutuality and women's economy to the formal marketplace. Wet nursing became a job in which the better-off hired lower-class women to live with the family for extended periods. Yet families tended to view wet nurses with suspicion. Increased class disparities led to "the growing cultural isolation of the working class" from middle- and upper-class families. As a result, wet nursing increasingly entailed the handing over of one's child to a stranger from a very different class and ethnic background and the product of an environment, the city, that middle-class families viewed as inherently unhealthy.[48] "The poor had their own districts, which by the 1840s were labeled slums and seen as well of poverty, social pathology, and disease. That the offspring of the well-to-do sometimes drank, literally, from such wells could be deeply troubling."[49] In general, American upper-class mothers were less likely than their European counterparts to hire a wet nurse and more likely to feed their infants cow's milk.[50]

Whether the statistics for infant cow's milk drinking in New York City are exaggerated or misstated, the fact remains that the rise of "the cult of true womanhood," accompanied by an ideology that encouraged, even glorified, breast-feeding, was accompanied by a rise in artificial feeding. Explaining this contradiction requires looking closely at the contradictory ideas surrounding the Romantic vision of motherhood.

The Domestic Mother and the Natural Child were constructed as an escape from "artifice," the Romantic term for the new, ruthless, competitive business world of industrial capitalism. In the language of the

Romantics, sentimental domesticity, childhood innocence, and sublime wilderness served as refuges from the rational, calculating, impersonal world of the city and the market. Each of these ideas stemmed from Rousseau, whose *Emile* was widely read as child-rearing advice in the early part of the century and whose ideas were taken up by the authors of the American child-rearing texts that began to appear in midcentury. In contrast to John Locke, who viewed children as a "blank slate" on which parents could write what they wanted, Rousseau viewed children as possessing innate and individual potentials that parents were responsible for nurturing. For Rousseau, childhood was a state apart from the adult world, a purer, more natural state of being that he advised parents to retain for their children as long as possible.

Similarly, Romantic ideas of domesticity had a significant influence on how men and women of the nineteenth century remade their relationships with each other in the context of industrialization and urbanization. Many historians have focused on the development of a separate "women's sphere" out of an earlier world in which public and private, rational and emotional, community and family, economic and familial, as well as men's and women's worlds, were one, intermixed and inseparable. This story shows that with the rise of industrialization and urbanization, farmers became city professionals or laborers, and fathers of both classes left home for work. The departure of the father from home led to the creation of public (work) and private (home) worlds. For working-class women, this often meant working at home alone in the presence of, and in partnership with, their children. For middle-class women, separate spheres meant a separation from economic production altogether.

The Romantics interpreted industrialization as the creation of the world of artifice out of a purer, more natural world. This resulted in a "radical redefinition of women's roles," which "glorified the American woman as purer than man, more given to sacrifice and service to others, and untainted by the competitive struggle for wealth and power."[51] Men, the rational side of the male-female duality since Plato, were overseers of a newly separate rational world while women, in their separate domestic sphere, oversaw the world of sentiment.

The "cult of true womanhood" became the social ideal, even for those without the resources to achieve it. "Although only a small proportion of women in industrial America actually were leisure ladies," Glenda Riley notes, "the concept was soon elevated to the status of ideology," giving women in aspiring up-and-coming families "a vehicle through which they might become, or at least emulate, ladies."[52]

The Romantics rejected both rational argument and animal urges for the sake of "higher" feelings, such as love, friendship, or feelings of the sublime in the presence of awe-inspiring natural landscapes. These feelings renewed male psyches and enabled them to return to their rational, competitive worlds. However, men mainly experienced these feelings within these worlds of refuge, and not in the public world, where rational thought and base instinct—such as competition and avarice—held sway.

Women were responsible for creating the domestic world of romantic refuge. Children, as natural innocents, added to the creation of this atmosphere. Some authors have argued that children became mere decoration in these urban bourgeois households. However, the Romantic idea of refuge and sentiment confounded symbol and reality, since children and women both represented and were considered responsible for the creation of the private sphere.

The private world of domestic refuge was also the place where citizens were made. Increasingly, the moral education required to create good citizens did not appear to be generally available in the male public sphere of the marketplace and corrupt politics. The middle-class wife became the "republican mother" responsible for the creation of a moral civil society.[53]

This contrasted significantly with the view of mothers in the colonial era. During this earlier period, "Americans did not idealize motherhood and often seemed unconcerned about it."[54] Women were viewed primarily as "helpmeets." Few parents referred to books to guide their parenting; child-rearing texts were rare. Interestingly, the few child-rearing texts that were written were addressed to fathers, since they tended to be at home. "Mothers, on the other hand, were so busy with their many chores that they often relied on their husbands or other adult members of the household to care for their children."[55]

Yet the idea of women as innately more loving, and the increasing placement of middle- and upper-class women in a noneconomic household sphere made women responsible for protecting not only the virtue of their children but the virtue and morality of the nation as well.[56] By the early nineteenth century, child-rearing texts addressed to women were one of the most popular forms of adult reading in America, along with sentimental novels.[57] While these books primarily influenced the middle class, they increasingly became the standards by which women charity workers judged their working-class clients.[58] Historians describing these texts today waver between treating them as "the way" women thought at the time and treating them as the way women were told to think. Either as accepted ideology or as an ideology to resist, the domestic republican mother became the standard for middle-class American women.

CONTRADICTIONS

A number of historians have noted that the cult of domestic motherhood was a contradictory ideal that put severe cross-pressures on upper- and middle-class wives. Middle-class women during this era were subject to "conflicting cultural demands" that separated them from each other and from their children as sources of assistance in the handling of household production and family care.[59] In addition, creating and maintaining the middle-class home put new demands on wives. For example, the social demands of visiting, entertaining, and making consumption decisions required a great deal of time. Unlike women's earlier roles, these errands and obligations were not ones in which children could tag along; the result was an increasing disparity between "what was considered truly important in adult society and what children were in fact capable of."[60] Visiting was one example of this type of social obligation and ritual that was "beyond the vocabulary of a child":

> "Fortunately," wrote Miss Leslie in 1864, "it is no longer fashionable for mothers to take their children with them on morning visits." Four years earlier, Florence Hartley had proclaimed, "To have a child constantly touching the parlor ornaments, balancing itself on the back of a chair, leaning from a window, or performing any of the tricks in which children excel is an annoyance, both to yourself and to your hostess."[61]

As a result, children were increasingly "relegated to—a special world apart."

> Cribs with guardrails and bars made bed a safer place than an unwatched cradle . . . varieties of toys and books enabled children to en-

> tertain themselves easily for substantial periods of time without active supervision. . . . [C]are that once had demanded physical presence could now be administered from a distance by means of substitutes for or extensions of personal contact.[62]

For fathers, "Adult recreation, with the exception of such sports as baseball and ice skating, was likely to take place in smoke-filled clubs, hunting expeditions, or other social functions at which children were unwelcome."[63] As a result, adults and children began to live in two different worlds. Yet, ironically, there began to arise during this period a sentimental idealization of children.[64] This process of "sentimentalization" made women and children the antidote for the new society in which public relationships denied mutual obligation. The family circle became the hearth of sentiment and mutuality that was both cherished and marginalized.

Along with this marginalization came an increasing isolation of the middle-class mother from the outside world. Increasingly, she lost her traditional forms of social support. Instead, she was asked to manage the family "in a society that rewarded independence and self-reliance."[65]

As de Tocqueville noted, many American women before marriage were given greater freedoms and were more independent than European women. They had greater access to education and greater control over determining their marriage partners. However, after marriage, American women appeared more isolated. Individual independence became the independence of the family unit, and the mother's role was taken to be self-sacrifice for the sake of her family and a self-reliance that drew the family away from its community.[66]

Harriet Martineau's description of the plight of the American woman was even more critical than de Tocqueville's. Martineau was also a visitor to the United States who wrote about the society she found there in the 1830s. She described the treatment of women in America as "below, not only their own democratic principles, but the practice of some parts of the Old World":

> While woman's intellect is confined, her morals crushed, her health ruined, her weaknesses encouraged, and her strength punished, she is told that her lot is cast in the paradise of women: and there is no country in the world where there is so much boasting of the "chivalrous"

> treatment she enjoys. That is to say,—she has the best place in stage-coaches: when there are not chairs enough for everybody, the gentlemen stand; she hears oratorical flourishes on public occasions about wives and home, and apostrophes to women; her husband's hair stands on end at the idea of her working, and he toils to indulge her with money: she has liberty to get her brain turned by religious excitements, that her attention may be diverted from morals, politics, and philosophy; and, especially, her morals are guarded by the strictest observance of propriety in her presence. In short, indulgence is given her as a substitute for justice.[67]

Martineau's major critique of American, and European, upper-class society was that it did not allow women to be useful. In such a context, the contradictory pulls of "duty" and "indulgence" must have been great. Certainly, breast-feeding was a context in which these contradictions were particularly evident. Women bred to be useless would most likely have difficulty managing the physical challenge of breast-feeding, which is tricky, sometimes debilitating, and often painful.

Yet this discussion of isolation tells only part of the story. Middle-class women of the mid-nineteenth century were not isolated in every way, they created new networks around another Romantic ideal: friendship.[68] Friendship involved the formation of intimate bonds revolving around the expression of personal, heartfelt emotions. However, these new networks, of friendship and feeling, entailed different forms of obligation than the earlier economic networks women had formed with kin and neighbors. While these women increasingly discovered communities of emotional self-expression, they lacked the earlier networks of "helping out" that were an integral part of an earlier women's economy. While women visited, wrote, and unburdened their hearts, they were also increasingly expected to be independent and self-reliant in the functioning of their households.[69]

The American middle-class urban woman was therefore caught between her role as the manager of an independent modern family, reliant on no one, and more traditional social expectations that she would fulfill and maintain a role that was based in an earlier form of community reliance. "The latent contradiction between woman's preparation for self-fulfillment and her role as the family's key nurturing figure often resulted in enormous personal tension, sometimes manifested in the classic nineteenth-century neurosis of hysteria."[70] Interestingly, this confined role for women also contradicted the political atmosphere of the time. The increasing patience and duty demanded of women in the home came at a time when Americans saw themselves as capable of great social achievements, when "change was considered a self-evident good, and when nothing was believed impossible to a determined free will, be it the conquest of a continent, the reform of society or the eternal salvation of all mankind."[71] For women caught in these clashing forces, moral reform crusades provided a crucial outlet.[72]

In New York City, this role would be taken primarily by the city moral reform societies, especially the associated city mission movement. The role of Robert Hartley in all of these movements brings the culture of the female body full circle: the context that created the Romantic mother—the physical and cultural restriction of women—also created a new rhetoric of milk as a perfect food that removed women from their physical role as food provider.

Ironically, the economic insecurity of working-class women in urban areas made them retain the "helping out" network. A male breadwinner's wages were not sufficient to feed and clothe a family. Therefore, other mothers and children were forced to gain an income of some sort. Kin also depended on pooled resources to make ends meet. "For many immigrants the kinship network provided day-to-day assistance with housing, child care, and loans."[73] In working-class neighborhoods, a form of "helping out" network often extended to neighbors as well. Therefore, while working-class women were increasingly uncertain about income, they existed in a larger support network that included a network of helping out.

For middle-class women, the "helping out" network must have seemed increasingly beneath their class status. As a result, the system of domestic work changed from one of mutual help to one in which women of one class worked as "domestics" for hire by women of another class.[74] In addition, many "hired hands" in the earlier system were relatives, such as children boarded with kin in order to help out, learn a trade, and receive further discipline.[75] The labor of working-class children became increasingly used in unskilled factory jobs, without the move away from home. For middle-class children, apprentice work away from home was replaced with play and education in the home, to develop their skills for the new professional world of work.

This new system, in which children remained at home with women while men spent most of the day away from home, made women responsible for the upbringing of children, not just ideologically but through their training in social skills. According to Priscilla Clement, "Because the job market was so much in flux, and new job skills necessary in an industrial economy might render learning traditional crafts useless for advancement, middle-class parents agonized about how best to rear and prepare their children for the future."[76] Children's upbringing and readiness for the professional world, rather than the property they were to inherit, became the main source of their future status. This led to a certain "status anxiety" in the middle class and made mothers increasingly important to the raising of children, particularly older children who had commonly left the household in earlier times, or who were expected to help with the younger children. In the new middle-class household, older children required education and could not be depended on as helpers.

In summary, the middle-class mother of the mid-nineteenth century was faced with the loss of a "community of practice": husband, kin, friends, neighbors, and older children as physically available help in child rearing. In addition, the increased demand on her time for the education and discipline of older children, as well as a growing number of demands from an adult world increasingly separate from children, had a great deal of influence on her decision whether or not to breast-feed her infants.

Not surprisingly, studies of contemporary influences on breast-feeding also note the importance of support networks for a mother to successfully perform this task, which "may influence not only a woman's decision to breast-feed, but also her feeling of success and satisfaction with the breast-feeding experience."[77] The drop in rates of breast-feeding that have accompanied the rise in urbanization and industrialization over the last century is only in part the result of aggressive advertising by infant formula companies or even the rise of women's participation in wage labor. For instance, Nomajoni Ntombela, the chair of the African regional office of the International Baby Food Action Network, links the drop in breast-feeding with the loss of social networks, particularly older relatives, who once taught young mothers how to breast-feed. "We make the mistake of believing breast-feeding is natural, an intuitive thing," she states. "But it's a learned behavior passed on from generation to generation. In the old days, the older women would sit there and encourage and tell you to do this and that—it was part of education."[78] Urbanization therefore does more than provide employment; it changes the kind of community a person lives in: who is looked up to, who is visible, who one interacts with, how one gains community esteem. As a result, the change to an urban way of life for Swazi women has meant a move away from the "community of practice" that relied on older women for systems of knowledge and that gained a person respect, esteem, and status through intergenerational and kin relationships. In contrast, urban communities of practice in Swaziland are based on peer recognition as a source of personal self-esteem; consumption and style are the ways to gain this status.

Urban women in nineteenth-century America were struggling with similar changes: urbanization and the changes it entailed. In both time periods, women were admonished to breast-feed. Yet the older community of practice that provided mutual assistance and made breast-feeding a source of self-esteem no longer existed, leaving women struggling to achieve urban-based cultural statuses while escaping the stigma of being "bad" mothers because they did not breast-feed. These women were caught between traditional and modern expectations. Because of their inability to meet both of these expectations at the same time, their societies increasingly looked upon them as less than perfect, as physically and psychically unable to cope with urban life. This representation of women as less than perfect only exacerbated the devaluing of their role as food provider within the family.

CONCLUSION

The introduction of cow's milk as an American food did not begin with a milk industry media blitz. Certainly, infant formula companies, once established, worked assiduously to increase the demand for this product. However, media promotion had nothing to do with the rise of cow's milk as an infant food in the 1840s. These larger-scale forces did not exist until at least two decades later.

The progress story trope of thinking by graph, however, creates a tendency to let the decades slip into one another, to meld the influences of the 1840s and the influences of the 1880s into one. When one looks at graphs that show the increase in milk consumption over time, it is easy to consider the food at the beginning of the graph as the same food at the end. It is also easy to see the forces creating the beginning of the line as the same, in miniature, as the forces creating the line today. However, this graph-induced thinking is problematic. Milk consumption in the 1840s is difficult to place on a timeline that ends with today's milk consumption. Yes, some fresh milk was consumed, but very little in today's mass, daily, raw, by-the-gallon state. The reasons for the rising consumer demand at the beginning of the graph—infant feeding—in fact no longer exist today. Most bottle-fed children consume a processed milk or soy product. Beyond bottle-fed children, milk drinking was a practice that occurred primarily under special circumstances: when a great deal of it was available, as during a spring flush on a farm, or on a special occasion.

Yet milk's early decades as an infant food already reinforced its unique position in American life: it needed to be produced every day of the year. Unlike the production of wheat, sugar, or tea, this daily production was local. Yet if the price rose, there was no substitute. The people who needed cow's milk—infants without breasts available to them—could not take it or leave it. The rise of milk drinking was therefore a process in which a food that needed to be produced every day for a specific group of consumers became an everyday food for everyone.

The only substitute was women's milk. Yet the breast milk production system was intrinsically problematic, given the change in social relationships that were a part of urban life. The informal community of practice that was part of the women's economy had broken down. It was this breakdown in an informal exchange relationship—and not the rise of an industry—that brought about the initial need for milk as a commodity. Yet this initial impetus would be rapidly replaced by new forces motivating new habits of milk drinking.

The explanation for why Americans began to demand a supply of milk from the countryside is not the explanation for why milk drinking became a major American food habit. The mechanism moving the line along the path of the graph changes through time. Consumer desires were part of the mechanism, which began with the desire to save babies. This desire was based on an identification of pathology—infant mortality—and the association of that pathology with cities. The route of the pathology was the woman's body, seen as ill-adapted to city life. Rather than seeing the breakdown of women's economic relationships, reformers represented the problem as the breakdown of women's bodies, due to their physical feebleness or moral degeneration, both of which were the result of city life.

In his book *The Pasteurization of France,* the French sociologist Bruno Latour examines the rise of hygienic medicine in France. He discovers that ideas of hygiene actually predate the discovery of the microbe. "The social context of a science is rarely made up of a context," he explains; "it is most of the time made up of a *previous* context."[79] One can say the same thing about the industrialization of food: it did not rise from the context in which it eventually existed, but from a previous context.

Latour calls the development of this context for the rise of microbiological hygiene the "pasteurization" of France. Pasteurization, for Latour, has little to do with milk, but with the microbial approach to hygiene. Yet clearly, the context of milk in American cities—which also eventually led to "pasteurization"—also had to do with problems of hygiene that were eventually located in microbes. The sharing of contexts between the two situations is therefore not surprising. Latour frames his discussion in semiotics—the study of signs—and shows how the working of signs set the stage for Pasteur's discoveries. In particular, he notes that the association of physical and moral degeneration with cities was a major theme of social commentators of the time. He calls this the discussion of "the conflict between wealth and health," which, he states, commentators saw as "reaching such a breaking point in the mid-century that wealth was threatened by bad health":

> The men, as everyone said constantly, were of poor quality. It could not go on like that. The cities could not go on being death chambers and cesspools, the poor being wretched, ignorant, bug-ridden, contagious vagabonds. The revival and extension of exploitation (or prosperity, if you prefer) required a better-educated population and clean, airy, rebuilt cities, with drains, fountains, schools, parks, gymnasiums, dispensaries, day nurseries.[80]

Hartley echoed the sentiment of the French hygienists. He attributed the need for country milk to the physical or moral degeneration of mothers who were unable or unwilling to breast-feed. The city was the source of this degeneration, through either overstimulation, disease, or temptation.

Yet, at the same time, Latour critiques this reductionist explanation, particularly the way this perspective leaves politics out of the causal equation. Harvey Levenstein makes a similar point when he questions whether lack of clean milk was the real problem for urban children at the time. "In fact, no one has yet been able to explain this century's decline in infant mortality in a manner convincing to most of the other experts in the field," he states.[81] Instead of attributing the lowering of infant mortality to the institution of public hygiene in cities, he emphasizes the role of rising incomes. In particular, he shows that differences in infant mortality rates between breast-fed and artificially fed children are significant, but not nearly as significant as differences in mortality rates between children in higher-income and children in lower-income groups, breast-fed or not. Levenstein therefore adds to Latour's skepticism about the argument between "wealth and health" that ignores increasing inequality between classes and focuses only on the city as an environment.

For many decades, milk was placed at the center of this depoliticized discussion about the impacts of an urbanizing society in the midst of rapid economic growth. It is easier to point to bad cow's milk than to increasing class inequality as a source of child mortality. And if milk was the problem, "reformers" could concentrate on reforming the food provision system rather than addressing the inequalities of the economic system.

The progress story of perfect milk therefore hides social and political facts by making questions about the rise of milk unaskable. Narratives of perfection, and their accompanying narratives of inevitability and enlightenment, throw a veil over social and political forces, the human choices, in history. The progress story is as apolitical as the notion of perfection itself.

4

The Milk Question

Perfecting Food as Urban Reform

AFTER HARTLEY'S ESSAY, it took four more decades for New York City to develop a country milk system. Even with the opening of new rail lines, 70 percent of New York City's milk continued to be from swill barns into the 1850s. State and municipal legislation in the 1860s and 1870s sought to prohibit milk adulteration and swill milk production, but with mixed success. In 1856 the Brooklyn Common Council passed the first law regulating the production of milk: an ordinance to restrict the number of cows on city lots. The large distillers quickly subverted the law by having their operations exempted from the new regulation. In his history of milk regulation, John Dillon called the exemption "probably the first 'joker' in an American milk law."[1]

A number of investigating committees and reports, such as a special report on milk adulteration by the New York Academy of Medicine, as well as journalistic exposés in the *New York Tribune* and *Frank Leslie's Illustrated Newspaper* joined the milk reform cries. Frank Leslie, an early muckraker, published a series of articles on the public threat of the swill milk trade in the 1850s, illustrated with sinister drawings of rough, male "milkmaids" milking dying cows in brewery stables (fig 4.1). For several years, Leslie continued this campaign in his newspaper and in the city courts.[2]

By 1880, railroads brought country milk to New York City from a wide area, including New York state, New Jersey, Connecticut, Massachusetts, and Vermont. Yet the arrival of country milk by train to New York City did not resolve the problems of milk safety and quality. If disease statistics are any indication, the railroad country milk system was not much of an improvement over swill dairies. One food historian, noting that urban infant mortality during this period was twice as high as the rate in the country, stated that "More milk had not meant pure

FIG. 4.1. Drawing of a dying cow being milked in a swill barn, *Frank Leslie's Illustrated Newspaper*. From Giblin, *Milk: The Fight for Purity*.

milk, and suffering from diseases carried by this and other foods—scarlet fever, diphtheria, and other ailments—was very great."[3] In addition, a number of summer epidemics late in the nineteenth century, and the suspected role of milk in spreading the epidemics, brought concern over milk safety to a head. Also, by the 1890s, experts began to argue that the specific form of tuberculosis in cows was in fact contagious to humans. This discovery increased the seriousness of the milk safety question.[4]

Milk reform was part of a broader food safety and city sanitation movement, which was itself part of the new "Progressive" politics of urban reform. City reform politics seemed to move seamlessly from a pietist to a Progressive agenda. The New York Association for Improving the Condition of the Poor (NYAICP), Robert Hartley's group, exemplified this transition. From its early roots in the evangelical mission movement, it became a pioneer in Progressive reform issues such as housing and public health. The genteel ladies and their "visitations" of families evolved into social workers and their casework methods. "In the development of contemporary attitudes toward poverty and institutions for its amelioration, the city mission generally, and the Association for Improving the Condition of the Poor particularly, played a significant role."[5] The idea that links both pietism and Progressivism is perfection, along with its opposite, degradation. The pious evangelical achieved perfection by discovering God's design. City missions uplifted the poor by saving them, since conversion led to a perfection of a way of life, which led to material success. Progressives identified perfection as the perfectibility of social institutions: regulation, welfare, science. The difference was a matter of authority: who had the right to name this perfection and make it real. The professionalization of the ideology of perfection took the impetus away from religious reformers and put it into the hands of experts. These experts identified the problem as food adulteration or "sophistication." For milk, this could include a number of chemicals, including formaldehyde, to retard spoilage, but more often involved the watering down of milk and the skimming of milk fat.

Ironically, it was the country source of milk that formed the problem. While Hartley spoke hopefully about the benefits of bringing pure country milk to the city, later food reformers noticed that the increase in distance between the producer and consumer gave many opportunities to adulterate food products. Transportation was a particular problem for milk: the time it took and the number of hands that milk passed through between the farm and the consumer led to both a deterioration and an adulteration of the product. The temptation to skim milk must have been great, since the cream could be made into butter or sold separately, thereby providing two incomes from one product. Unlike today's struggles to lower fat consumption, the nineteenth-century city and its countryside struggled over access to the butterfat in milk. In the earlier system, farmers consciously and openly sold the by-product of their buttermaking to the city, often in a fermented state. But the primary consumers of sweet milk, weaned infants, could not survive on buttermilk. Infants needed the fat in milk to thrive. The resulting struggle was a fight between farmers and urban parents for the right to the butterfat, with the "iron cow" and the city inspector playing important roles in the struggle.

This new concern over the safety of food provisioning systems involved a reformulation of city-country relationships and a new view of the role of country people in the nation. By World War I, America had experienced a rupture in the moral linkages between rural and urban people. While many urbanites continued to see the farmer as "the prototypical American, the independent, self-reliant, natural, productive, middle-class yeoman, the rock of republican government and the conservator of national morals," a growing number of city people viewed the farmer as "the worst in American Society. To them he epitomized the crudeness, waste, ignorance, and degeneracy of society. He also symbolized the worst of the selfish, destructive, unrestrained individualism with which many Europeans associated the United States."[6] Even as early as 1884, the muckraker Henry Demarest Lloyd, writing in the popular *North American Review,* took to task a group of three hundred Orange County farmers who were carrying out the country's first "milk strike": the withholding of milk from the market for a higher price. He described these farmers as a "trust," in the same league as the railroad barons and various steel and other manufacturing combinations: "The trade in milk at the point of largest consumption in the United States now rests in the hands of these combinations."[7] This lack of distinction between robber barons and farmers was not unusual by this time. Many believed that "[j]ust as the unrestrained individual in business, labor, and politics had to be checked in the interest of national progress . . . so too did the unrestrained individual on the farm."[8]

Even before the 1920s, when rural people became the political minority, American politics had moved away from the idea that farmers were the productive base on which the wealth of America stood. Yet many people, especially rural people, remained "agricultural fundamentalists." They held to the idea that what was fair to the farmer was most beneficial to the country as a whole, and that was to assure the farmer a price that returned a reasonable profit.[9] However, with the growth of a more industrial America at the end of the nineteenth century, it seemed to many that the capitalist entrepreneur, not the producing farmer or laborer, was the source of the country's wealth. The welfare of the farmer was no longer seen as the key to the general welfare. Increasingly, urban publics proclaimed that the farmers' role was primarily to provide inexpensive food to the urban populace.[10]

The rise in prices during the first two decades of the twentieth century was one major factor that turned city people away from the idea of farmers as the core of national wealth and the country as a source of moral sustenance. "The rising cost of living was quite disturbing to a generation of consumers which grew up in an age of static and even declining prices, . . . but the wholesale price of farm products increased at nearly three times the rate at which the general price level rose."[11] A public that once celebrated the farmer as the backbone of the nation began to turn away as the price of food climbed:

> The high farm prices were deeply resented by the consumer public. Sometimes farmers were denounced as conscienceless monopolists who set the prices of the necessities of life to the disadvantage of every city dweller. Manufacturers claimed that the American farmer was lazy and inefficient. If only he would get busy and increase production, food would be cheaper, the wages of city laborers could be lowered, and the American manufacturer could the better meet foreign competition. But the farmers were disturbed only by the fear that the good prices might not last.[12]

Concern for farmers moved from a cooperative view of city and country moving forward together to a concern for the country's ability to feed the city. By the New Deal, this view of the countryside had become most prominent, at least among city people. Roosevelt's declaration that "the welfare of the farmer is of vital consequence to the welfare of the whole community" was based not on the producer's importance in sustaining the nation's wealth, but on his ability to supply abundant food for the cities.[13]

Farmers also did not fully escape blame in the controversies over food adulteration. Dairy farmers in particular were accused of being first in line to use "the cow with the iron tail," with the water pumps of milk dealers next in line.[14] The discovery and eventual acceptance of the germ theory of disease late in the nineteenth century brought farmers under even more suspicion. In no case was this more evident than in the case of milk. A typhoid epidemic in Washington, traced to the city's milk supply, prompted Theodore Roosevelt to authorize an investigation of the milk problem. The result was the study *Milk and Its Relation to the Public Health.* Published in 1908 under the direction of the Public Health Service Hygienic Laboratory director, Dr. Milton Rosenau, the study investigated every aspect of the industry and the various ways suggested to resolve the milk problem. These investigations established milk's ability to carry a wide range of contagious diseases, including typhoid.[15] Typhoid carriers could survive with the disease, sometimes even unknown to themselves, yet the infants and children who consumed the milk risked death. This fact made the urban public even more fearful of countryside production. Pictures of unsanitary barn conditions circulated in published material and led to the establishment of dairy farm inspection systems in many states. The urban food reform movements that had welcomed the country farmer now suspected this source of milk as well.

The reports seldom made the link between the new system of market milk dairy farming and contagion in the milk. However, the type of farming required to fulfill the daily urban need for fresh, fluid milk was itself more likely to increase health risks. Farmers supplying this "market milk" generally milked larger herds that spent more time confined to the barn. The data on the geographic extent of bovine tuberculosis at the time provide an indication of the problem: inspection of herds for tuberculosis found that only about 1 percent of herds that spent most of their time out of doors were infected. Herds that spent most of their time in stables had a significantly higher degree of infection.[16] For cows, no less than people, distance functioned as a form of sanitation. Large, stabled herds were necessary for the creation of a year-round market milk system.[17] In other words, the "country" dairy farm that supplied milk to cities, while certainly cleaner than swill milk dairies in the cities, and certainly providing cows a more varied feed ration, shared some problems with swill dairies in the way cows were fed and managed. The large-scale daily provision of fresh milk to cities required a particular dairy farm strategy, but that strategy increased the riskiness of the supply.

By the late nineteenth century, the concern about milk had turned from adulteration to bacteria. While Pasteur and Koch made their discoveries about bacteriology and contagion several decades earlier, it was not until 1890 that the idea of bacterial contamination as a problem began to take hold.[18] Food reformers, who had achieved some success in lessening the adulteration and watering of milk, now turned to advocating ways to rid milk of dangerous bacteria.

The people who lobbied for pure milk in New York City in the late nineteenth century were different from midcentury reformers. Many were upper class philanthropists, like R. H. Macy's owner, Nathan

Straus. Straus opened several pure milk stations in the summer months for poor children and to advertise the benefits of pasteurization. Through these stations and many books and pamphlets, Straus campaigned assiduously in favor of pasteurization of the entire milk supply.[19]

As a result, pure milk reform movements turned away from the dairy farmer as the salvation of the urban populace to the idea that the farmer himself must be put under urban control. As a result, New York City and other cities began to enact legislation targeted at reforming the dairy business, both dealers and farmers.[20] Unlike the ministers and merchants of the mid-nineteenth century, the new milk reformers were members of what Robert Wiebe has called the "new middle class"—scientists, doctors, industrialist philanthropists, and government officials.[21]

From the 1850s on, the need for public regulation of city food supplies was a continual topic of debate, and this debate focused on one food in particular, milk.[22] This movement reached a high pitch in the 1890s, when local philanthropists, such as Straus, created their own systems of milk provision for children. Consumers waged pure food movements in other cities as well, such as Milwaukee and Providence.[23]

Historians have noted the decline in rural and agrarian political power as populations became more urban.[24] New York state, as the first state to urbanize while retaining a significant rural population, led the way in this political transformation, giving urban groups a substantial say over the nature of agriculture in the state long before such shifts in the balance of power in other states. In this way, New York state's urban/rural political shift prefigured the power balance of the nation as the national demographics repeated this pattern. In other words, as New York City grew as a demographic and political force, it was able to gain more control over the countryside that provided a substantial portion of its food.[25] As a result, to a greater extent than any other state at the time, issues that affected the New York state farmer were worked out in New York City, in places like the conferences and meetings of the New York Milk Committee. These urban interests characterized the issue as "the milk problem" or "the milk question."

BUT WHAT EXACTLY WAS "THE MILK QUESTION"?

One major figure involved in this issue was the noted medical director of the U.S. Public Health Service's Hygiene Laboratory, Dr. Milton

Rosenau. Rosenau's book *The Milk Question*, published in 1912, provided the arguments in favor of pasteurization the same year New York mandated pasteurization of its milk supply. The primary purpose of Rosenau's book was to publicize his research demonstrating that the rapid heating method of pasteurization rendered milk safe from bacterial contamination. However, Rosenau also used the occasion to state his broader ideas about how milk industry organization and regulation could help resolve the milk question, which he defined in this way:

> Why do we have a milk question? Why all this fuss about milk and milk products? We do not have a bread question, a grain question, a fruit question, or a vegetable question—these substances also represent standard articles of diet. The answer is simple. We have a milk question because *milk is apt to be dangerous to health.* In fact, the milk question as we understand it to-day began only when it was shown that impure milk is apt to convey disease. This alone would be sufficient reason, but in addition we have several other important facts. One is that we cannot do without milk. It is true that several large nations comprising millions of people get along reasonably well without the use of the milk of the cow or any of our mammalian friends—rather our domesticated slaves. Western civilization, however, has come to depend upon cow's milk as an essential article of diet for children and it has become a very important article of diet for adults; it is therefore no overstatement for us to say that milk is a necessary article of diet.[26]

Unlike the earlier milk reformers, Rosenau does not claim that milk drinking is universal or even primal. However, he does insist that milk is a food Western civilization "cannot do without." American dependence on milk in the diet had moved from God's Design to national identity. The milk question was therefore embedded in a contradiction: milk was both essential to the American public for its health and harmful to its health. The question was, how do we provide ourselves with this necessary food without being poisoned by it?

The Conference on Milk Problems convened by the New York Milk Committee in 1910 illustrates the politics of the milk question. The voices heard at the conference show how the genuinely problematic character of milk was constructed in particular ways. The New York Milk Committee was not an unbiased group; heavily influenced by physicians, it had lobbied against the pasteurization mandate for many years. The conference participants were therefore heavily weighted against pasteurization. Looked at in conjunction with Rosenau's book, the conference proceedings give the full gamut of political opinion and activism on milk at that time.

CONVENING ON MILK PROBLEMS

In turn-of-the-century New York City, the state of milk politics had changed substantially from the time of the pietist antebellum reformers. Yet the New York Milk Committee also represented continuity with past city food reform. This committee was formed by the New York Association for the Improvement of the Condition of the Poor, the group founded by Robert Hartley more than half a century before. The New York Milk Committee's Conference on Milk Problems occurred at a particularly important time: within a year, New York City would decide to require the pasteurization of most market milk.[27] This was at a time when many different arguments about how to organize the milk industry came to a head. In many ways, this particular conference represented a culmination of seventy years of discussion about milk.

One major indication of the change in the politics of milk reform is the list of participants in the conference. Neither milk dealers nor farmers testified. City doctors, their organizations, and state officials provided most of the testimony. Unlike activists in the earlier period of milk reform, speakers on the subject were neither ministers (like Hartley), nor journalists (like Leslie), nor philanthropists (like Straus). The participants in this discussion were not the private citizen reformers who fought against swill dairies, but public experts hired to present scientific information.

In the welcoming address to the conference, the secretary of the committee defined the milk problem as

> one of the most important problems that confronts this community, and all the metropolitan communities of this country. Tersely put, I think it can be said to be this: How can we so protect the milk supply as to make intestinal diseases of infants fed with cows' milk, as rare as smallpox?

Noting that "we are beset by the exponents of various plans for furnishing clean and safe milk," he laid out about two dozen questions the conference was to address, including "What injury, if any, results from Pasteurized milk?" "Will Pasteurization make unnecessary the inspection of milk shops, depots or country dairies and creameries?" and "If economies in the distribution of milk are possible, and will lower the price of milk to the consumer, what steps can be taken to bring such economies about?"[28]

The speaker then asked the conference to address one final question: "Is it desirable and feasible to provide some form of maternity insurance, enabling women who are engaged in industrial employment to refrain from such work for sufficient time prior to and after childbirth?" This question implied a program that would assist women to stay home and breast-feed their children. Interestingly, no speaker at the conference addressed this question.[29]

These reformers no longer had to exhort farmers to sell milk to the city. Officials at the conference testified that over forty-thousand farmers in six states, and as far away as four hundred miles, were selling milk to New York City by this time. The single private industry official to testify at the conference, a representative from the Lackawanna Railroad, gave substantial evidence of this fact. He noted that New York City derived its entire milk supply by railroad, most of it two hundred to four hundred miles away.

There was a great deal of contention about pasteurization in the conference testimony. Despite the fact that the conference was held on the eve of New York's pasteurization mandate, the committee testimony is remarkably indecisive about pasteurization. While the committee was unusually sympathetic to the anti-pasteurization lobby, the testimony also indicated that the committee was truly "beset" by supporters of two different ways of protecting milk safety: the certified milk system and pasteurization.

The committee was heavily influenced by city doctors who supported the idea of "certified" milk. As knowledge about bacteria and sanitation grew, city doctors had begun to focus more on the sanitation problems of the city, including sewer, water, and milk provision systems. Two alternative systems had emerged to deal with the milk problem: the certified dairy and the pasteurized milk system. The certified dairy system, begun by the pediatrician Henry Coit in New York in 1890, was a labor- and capital-intensive form of dairying in which milk was produced under the direction of a city Medical Milk Commission, composed mostly of doctors. Cows were frequently inspected and milking practices and barn conditions were exemplary, as "certified" by a medical board. Certified dairies were generally the largest dairies in any state at this time, with many operations milking over five hundred cows.[30] While some celebrated certified milk as fostering "a revolution in the methods of producing milk and the method of its supply to large cities," its high price prevented most but the very wealthy from purchasing this superior product.[31]

The second way to make sure of the safety of milk was through pasteurization, a process Pasteur invented in the 1860s, although not used to purify milk until the 1890s, when Sheffield Farms Dairy—later part of Sealtest/Kraft—installed the first pasteurizer in a New Jersey plant.[32] Many of the larger dairy companies therefore were already pasteurizing their milk by the time the city passed its pasteurization requirements.[33]

This distinction between the "raw" and the "cooked" was not lost on the conference speakers. Speakers who supported the feeding of raw, certified milk to babies tended to be private medical experts, specifically pediatricians. In general, raw milk supporters argued that heating destroyed many of the nutritious properties of milk, as well as the beneficial bacteria.[34] The pediatrician Arthur Meigs, for example, stated in his 1896 textbook on infant feeding that

> The sterilization of milk has been vaunted very much in recent years. . . . But latterly there seems to be a tendency to the general agreement that so high a temperature produces changes which render the milk less desirable as a food for infants, if it is not put in a condition to be positively injurious.[35]

In addition, as a professional group, pediatricians were concerned with interference in their customization of milk formulas for their patients. Infant feeding advice played a significant role in the rise of pediatrics as a medical specialty, resulting in the "'medicalization' of a set of activities that had been the responsibility of mothers and nursemaids."[36] New York City was the center of these new ideas. Two of the founders of American pediatrics, Abraham Jacobi and Luther Emmet Holt, were New York City practitioners who, through their concern for infant nutrition, became involved in the city's milk politics.

The extent to which pediatricians built up their expert role as infant feeding advisors simply in order to raise the status of their specialty is a matter of debate. "[N]evertheless, the professional needs of pediatrics were influential not only in highlighting how crucial was infant feeding but also in shaping the way in which the improvement of such feeding was conceptualized and pursued."[37] In particular, pediatricians tended to focus more on infant feeding rather than on breast-feeding. Turn-of-the-century physicians echoed the concerns expressed by mid-nineteenth-century milk reformers: that women's physical, moral, and emotional states, as well as the pressures of modern life, could affect the quality of their milk.[38] While the problem—women's bodies—remained the same between the religious and expert reform periods, the solution changed from a reliance on country purity to a reliance on urban expert knowledge.

As a result, those who testified in support of pasteurized (or, as one official called it, "cooked") milk tended to be public health officials or private medical experts. The Conference on Milk Problems heard from nearly a score of these experts, from New York City, other large and small northeastern cities, and Canada. Public experts testified that government was not capable of guaranteeing the safety of milk through inspection; pasteurization was necessary. Private experts threw doubt on the ability of technology and industry to solve the milk problem. They testified on behalf of their own private inspection system.

In 1906 New York City had mandated the inspection of market milk farms serving the city. At the conference, state officials reported that thirty-three city inspectors were assigned to inspect the forty-four thousand farms in six states as far away as four hundred miles from the city. They testified that this force was not sufficient to assure the safety of the milk supply. Yet the more rigorous inspection of certified milk put the price of this product out of the range of most consumers. As a result, certified milk represented only 1 percent of the total milk supply to the city.[39] While most of the government officials testifying agreed that certified milk was a higher-quality product, most insisted that pasteurization was the next best alternative. As one official stated, it was the state's duty "to make this vast milk supply *safe*, and to protect it from these many factors of contamination by careful and perfect pasteurization."[40]

Dr. Charles North, secretary of the New York Milk Committee, agreed. "The appropriation for milk inspection and the force of milk inspectors at the disposition of our city Department of Health has been insufficient,"[41] he stated, reporting that the New York State Department of Agriculture "does its best with its appropriation of somewhat over twenty-three thousand dollars for dairy purposes, and a staff of state milk inspectors which varies from eight to twelve."[42] He argued that three times as many country milk inspectors were necessary to inspect milk on the farm and twice as many were needed to inspect the twelve thousand milk outlets in the city.

The battle between pasteurization and certification was not simply an argument between pediatricians and state officials concerning authority. The critical question concerning certified milk was whether or not the state would attempt to replicate the certified milk system, with its labor-intensive inspection strategies, to guarantee the public milk supply in the same way that it guaranteed certified milk. As North described it,

> The raw milk supplied to the consumer of New York at the present time is a mystery. It carries no label to indicate whether it is good or bad. The public is in the dark. They have the right to assume that since the Department of Health issues permits and allows this milk to be delivered at the door, it is safe. Certified milk and guaranteed milk carry their credentials on the bottle, and are endorsed by the Health Department as safe.[43]

The state's role as licensor of milk dealers, according to North, made the state responsible for guaranteeing the safety of the milk they sold. In other words, the state license to sell milk was an implicit certification of milk. From North's point of view, the right to issue these licenses or "permits" implied a contract between the state and its consumers to guarantee the safety of the milk supplied. The state could fulfill this duty in two ways: (1) expand its inspection force to give the public milk with the same guarantees of safety as certified milk, or (2) choose the path of mandatory pasteurization (while maintaining some inspection), which would be a way to extend those guarantees to the general milk supply without expending the extra public funds. Private companies, particularly larger companies, through their capital investment in pasteurizing technology, would enable the state to supply the guarantee of milk safety without imposing further public costs.

But pasteurization, it was quickly recognized, was not the entire solution. In fact, the earliest accounts of pasteurization describe problems with companies pasteurizing and repasteurizing old and unsanitary milk to make it salable.[44] The continuation of some state inspection would provide an indication to consumers that milk was not being pasteurized just to cover up its defects.

The choosing of large-scale technical systems, therefore, was not simply the invisible hand at work increasing economies of scale and efficiency of production. The technical system was actively chosen by state officials as a solution to a number of political problems they faced, problems not entirely defined by politically neutral terms of "efficiency." The question in this case was not simply costs and benefits, but who paid for the costs, namely, taxpayers through an inspection system or private firms through investments in new technologies. Whether or not the private companies could pass on this cost to consumers had to do with what economists call "market power." As the next section will show, the capital costs of meeting pasteurization requirements made the milk industry increasingly oligopolistic and therefore increasingly able to pass on these costs.

Pasteurization enabled another group to avoid increased costs: farmers. A system closer to certified milk would have required all cows to be tested for bovine tuberculosis. Farmers balked at the increased costs of this potential requirement. The conference voted on several resolutions, one of which was that all cows be tested for tuberculosis prior to sale. During the comment period, dealers and producers in the audience stood up to speak against this resolution. One dealer remarked, after hearing the resolution read, "That would be a good thing to pass if you want us all to go out of the milk business."[45] The president of the Western New York Milk Producers' Association stood up from the audience to describe the implications of the resolution:

> In the last few years cows have cost us an advance of from $40 to $80. You test these cows and they will cost us from $100 to $150 and cannot be had. You test the cows of the State of New York to-morrow and only those cattle to be put into the dairies that pass the tuberculin test, and there aren't cows enough in the State of New York to produce your milk. And if you think the farmers of the State of New York and the farmers of the other states are fools enough to go on and do it merely *pro bono publico*, you are mistaken.[46]

Milton Rosenau agreed that farmers, if they were to make cleaner milk, should receive higher prices for that milk. "The farmer is not a philanthropist, but a business man," he stated. "His enterprise must be an economic success or it will prove a sanitary failure."[47] Echoing the milk reformers' voices seventy years before, Rosenau argued that consumers needed to be willing to pay the prices necessary to produce pure milk.

The tuberculosis testing resolution did not pass. For farmers, consumers, government officials, and industrialists, pasteurization was a politically more feasible alternative than universal tuberculin testing. Farmers were unwilling to pay the costs of this testing without threatening the political agitation necessary to pass these costs on in the form of a rise in the price they received for milk.

Was pasteurization a bad thing? Probably not. Even today, with universal tuberculin testing and strict bacteria standards, pasteurization protects the public from contamination by contact with contagious milk handlers and from an intensive production system that makes animal infection a common occurrence. Pasteurization does tend to destroy vitamin C in milk, but milk is not the major source of vitamin C in the contemporary American diet. Certainly, when it was used as an infant food, this fact was important. Despite these various doubts, the U.S. Public Health Service studies linking typhoid transmission and milk delivery certainly argue the benefits of pasteurization. In fact, typhoid cases dropped markedly after mandatory pasteurization laws were passed (see table 4.1).

Table 4.1
Typhoid Fever Deaths in New York City

Year	Population	Deaths	Rate per 100,000
1870	1,340,704	533	40
1880	1,777,351	443	25
1890	2,420,817	534	22
1900	3,225,324	673	21
1910	4,785,190	558	12
1920	5,683,765	137	2
1930	6,962,305	62	1

Source: Park, "The Relation of Milk to Public Health," 203.

In other words, the progress narrative makes some good points in favor of pasteurization, points that have been more convincing to the public, over time, than the arguments of those who continue to advocate raw milk. Yet the effects of pasteurization went beyond the elimination of dangerous bacteria in milk.

THE SOLUTION: THE INDUSTRIAL VISION

The advent of pasteurization had an enormous effect on the organization of milk processing and distribution. The economists Spencer and Blanford describe the structural impact on the industry: "The primary purpose of pasteurization, of course, is to reduce to a minimum the risk of disease transportation through milk. However, pasteurization also enhances the keeping quality of milk. It thereby facilitates large-volume operations and the distribution of milk over extensive areas from central plants."[48] Pasteurization therefore enabled larger companies to increase their economies of scale by collecting milk through a centralized plant from a larger area of procurement. As a result, the freshness and purity of milk became the product of an industrial system. With adequate investment in facilities, milk companies could sell pasteurized milk much more cheaply than certified milk.

Pasteurization mandates in cities led to significant shakeouts in the number of firms supplying those cities with milk. For example, in the Milwaukee milk market, the number of milk distributors decreased from two hundred to thirty-two between 1914 (when the city pasteurization ordinance went into effect) and 1920.[49] A study of plants in the Detroit area showed that the number went from 159 to 68 after pasteurization became compulsory in 1915.[50] In other words, the implementation of pasteurization regulations transformed an industry of small-scale dealers into one of major, large-scale, capital-intensive distributors.

Not long after this shakeout, the dairy industry began a period of intensive merger activity, creating two major dairy companies that dominated many of the eastern markets: Borden's and National Dairy. Beginning in the 1920s, these companies initiated expansion policies involving the purchase of hundreds of smaller companies (table 4.2). By 1932 these two companies purchased one-third of all milk in the north Atlantic states and over 20 percent of all milk in the United States as a whole.[51] Borden's handled about 20 percent of milk in the Chicago market and was active in many other markets across the country.

Table 4.2
Mergers of Major Milk Companies in the New York City Milkshed, 1923–1940

Year	Number of Mergers *National Dairy*	*Borden's*
1923	3	1
1924	9	0
1925	31	1
1926	6	1
1927	4	6
1928	62	25
1929	98	120
1930	71	40
1931	23	14
1932	3	21
1933	10	12
1934	10	5
1935	10	21
1936	23	40
1937	30	33
1938	22	22
1939	42	22
1940	34	21

Source: National Commission on Food Marketing, *Organization and Competition in the Dairy Industry.*

Gail Borden entered the milk business as an inventor of a process to condense milk, opening his first milk condensery in the 1850s. Borden was also one of the first to utilize continuous-process canning technologies. While condensed milk went over better than Borden's earlier invention, the "meat biscuit," he had difficulty creating demand large enough to justify the large capital investment necessary to support this technologically intensive production system. However, both the Civil War and World War I created a market for this product as a ration for soldiers. This turned out to be enormously profitable for all the condensery companies.[52] Companies that invested in expensive continuous-process machinery needed to guarantee themselves a large and continuous market for the heavy capital investment and high production that resulted from that investment. As a result, these companies also tended to innovate in two other areas: mass advertising and mass distribution. Borden's was one of the earliest companies to take advantage of continuous-process machinery. The combination of continuous-process technologies, a business organization that divided production among subdivisions, mass distribution, and mass advertising made Borden's a "modern" industrial corporation.[53]

Officials like Rosenau recognized that the increase in sanitary requirements on the farm and pasteurization in processing plants would significantly change the structure of the dairy industry. He and others concerned with milk safety, in fact, welcomed this change, arguing that only larger-scale firms could produce safe milk for city markets. Not only larger firms, but larger farms were necessary in order to make milk pure. "One of the real sources of trouble in the milk industry," Rosenau claimed, "is that the bulk of the milk comes from the small farm, and is there regarded only as a by-product."[54] If these small farms were eliminated from the milk market, the product would be improved:

> It is the opinion of many persons who have given the question thoughtful consideration, and whose judgment is worthy of respect, that the day of the small dairy man is doomed; that the production of milk will gradually and inevitably drift into the hands of larger dairies where economic conditions justify competent assistants, skilled supervision, and efficient equipment. In other words, the dairy industry is a special industry requiring technical skill of a high order and must become a specialty like other trades and professions.[55]

Rosenau welcomed the industrial model of farming as the road to progress for the consumer. He fully understood the social impacts of this change on farmers:

> The crowding-out of the small farmer is not to be lightly regarded, for he has human rights and society must grant him economic justice, but the crossing of the roads has been reached and the sign-posts are plain—either the farmer must comply with the exactions of the sanitarian or his milk will soon find no market.[56]

Rosenau and others argued that the investments necessary to make sanitary milk were not high. However, the description of the sanitary operation—with its assistants and equipment—is in contradiction to his argument that "by the exercise of cleanliness and a little ice" any farmer could meet sanitation requirements.[57]

Rosenau also welcomed the reorganization of the dairy "middlemen" into a few firms:

> The middleman, or large contractor, is a power for great good in the milk industry. In fact, this concentration of the business is just the right key which fits the lock of many of the most perplexing problems in our subject. . . . It is evidently much easier to control, educate, and regulate a few large contractors than hundreds of small independent dealers.[58]

An illustration from Rosenau's book *The Milk Question* vividly portrays the industrial vision of milk provision as superior to the small dealer system. The illustration follows milk through the inept and politically shady system and through the ideal, industrial system with clean, orderly employees and rational production, inspection, and transport methods (fig. 4.2). The picture implies that only the modern, industrial system could deliver milk safely to the city.

Other milk officials echoed Rosenau's point of view. Julius Moldenhawer, a state Agriculture Department officer, in his testimony at the Conference on Milk Problems, faulted the media for criticizing large milk companies:

> All are blaming the so-called "milk trusts" for every kind of trouble, while after all the consumer receives a better and safer product to-day from a number of large companies, than when the milk was handled by thousands of small dealers. *A properly controlled consolidation of the city milk supply is of equal benefit to producers and consumers.*[59]

In other words, the answer to "the milk question" was the rise of an industrial dairy-industry as a force for the benefit of the consumer. By the turn of the century, officials and experts turned to large-scale, capital-intensive, modern corporations and larger, more intensive farmers as the way to get milk to the cities. From their point of view, large corporations and larger farmers were the actors that could complete milk's story of progress, a story that had been told for over seventy years.

This solution to the safety question increased the power of the larger dealers. They could afford not only the pasteurizing plants but also the testing and other extra expenditures to assure the quality of their milk. Borden's, which had instituted pasteurization before it became mandatory, presented itself to the public as *the* safe milk company. State officials in the milk industry reinforced this view. Norton and Spencer, dairy agricultural economists at Cornell, claimed in 1925 that

FIG. 4.2. Educational placard from the Philadelphia Milk Show, 1911. From Rosenau, *The Milk Question*.

"New York City is now said to have the best milk supply of any large city in the world." However, they qualified this statement by saying, "The milk delivered by the larger milk-distributing organizations, at least, is uniformly pure, fresh, and of good keeping quality. This was not true twenty years ago."[60]

A COMPARISON WITH OTHER STATES

Whereas in other states, farmers still represented a substantial political force, the growing power of urban constituencies in New York City gave the city a great deal of say over the nature of agriculture in the state. While Chicago, Boston, and some southern cities had urban publics that were vocal about the nature of their local milk industry, only in New York City did milk issues so dominate the political scene. For example, two cities in the Wisconsin milkshed, Milwaukee and Chicago, campaigned for a pure milk supply. Yet in Milwaukee, the health department and private hospitals initiated and carried out the pure milk campaign, often despite public apathy and even resistance to the price increases that accompanied increased milk inspection.[61] Also, in states with a more powerful agricultural constituency, milk producers were well organized to resist health department regulations. Wisconsin farmers around Milwaukee took the city tuberculosis-testing ordinance all the way to the Supreme Court.[62] In Chicago "the precise alignment of science, business, and politics" that had been necessary for the enactment of milk purity regulations in other cities "broke down."[63]

In Milwaukee, health officials were not able to institute the milk controls enacted in eastern cities. One historian attributes this difference to a lack of political astuteness on the part of these midwestern officials.[64] Yet the history of milk safety movements in both Milwaukee and Chicago also reflects the general weakness of organized urban political support and the continuing strength of rural interests in those states. The continuing power of rural citizens in the Wisconsin and Illinois state legislatures contributed to the cities' struggle to regulate the milk industry.[65]

While descriptions of pure milk campaigns are prominent in histories of the Progressive and public health movements in other cities and states, histories of the Progressive movement in the West are more reticent on food issues. In California cities, for example, consumer food issues did not appear as often, or with as great a ferocity. While Progressive movement reformers fought for pure milk and unadulterated food in eastern, midwestern, and southern cities, the Progressive movement in the West focused on other issues. For example, a survey of the *Los Angeles Times* during the 1930s reveals few articles on milk safety or prices, a period when the *New York Times* printed "one news story on milk every second day and an editorial every two months."[66]

CONCLUSION

The decision to mandate pasteurization was therefore tantamount to an "industrial bargain" between farmers, consumers, and large industrialists that enabled public officials to avoid the difficult path of imposing costs on these politically powerful groups. However, looking at pasteurization as part of an "industrial bargain"—as a way that urban and certain rural political interests determined the large-scale structure of an industry—leads to some important questions that simple progress versus downfall stories of pasteurization ignore.

As the following chapters will show, mandatory pasteurization laws created an industry structure that eventually created more social problems. The political struggles between farmers, consumers, and large-scale dairy companies led to further state-sponsored compromises that created even greater dependence on fluid milk markets over other forms of dairying. After the mandatory pasteurization decision, government officials increasingly depended on the newly established but growing large dairy companies that not only willingly pasteurized their milk but also put in place their own private inspection systems. These companies established a set of company sanitation standards for the dairy farms that sold milk to them, before these standards became mandatory. Increased industry scale, which was bound to result from the pasteurization mandate, was seen by many as a positive change in the industry.

In New York this combination of safety, year-round reliability, and price considerations made state officials turn to large companies, large cooperatives, and larger farmers as the solution to the provision of milk to the urban populace. Because an industry structure built of smaller milk companies and smaller farms would require a more expensive public inspection system, this option received little consideration from government officials and academics.

Recently, we have witnessed the rise of a consumer politics that rejects the industrial food system in favor of smaller farmers and alternative food companies. Yet the rise of food consumers as a political force was part of the initial rise of a politics in favor of industrial food. In New York state, where consumers held significant and growing political power, government officials favored the growth of an industrial food structure that would enable cities to be fed cheaply and safely. The political relationship between city and country moved from the previous agrarian ideology to an urban-industrial one. Under these political circumstances, smaller farmers had no voice. They were too geographically and politically isolated to have much influence in New York dairy politics.

These are the essential elements of what I will call the "industrial bargain": an alliance between consumers, mass-production capitalists, and intensive farmers in the creation of a system of cheap nutrition. The industrial bargain was basically an anti-agrarian politics that saw farmers as servants of the city, and no longer as political actors with their own contribution to the nation.

5

Perfect Food, Perfect Bodies

IN THE 1880s, a pictured advertisement began to appear in women's magazines (fig. 5.1). It was one of the first magazine advertisements to include an image of any sort. The ad shows a small nest of baby birds being fed by their mother. Under the heading "Nestle's Food" came a number of testimonials and citations from medical books celebrating the product's ability to successfully feed children, especially those suffering from cholera infantum, the most prevalent, and mortal, commonly milk-borne disease of infants at the time. Though the ad focuses on the treatment of ill children, toward the end it recommends Nestle's "where the mother's milk is insufficient" as well.

The magazine ads that begin to appear, especially in women's magazines, shortly after the Civil War provide a substantial record of what Americans have wanted and dreamed for themselves for the last century. The images seem quaint to us today, but at the time they represented the new ways Americans were thinking about themselves. The novelty of these images becomes particularly apparent if one examines the previous images Americans used to represent their food system.

Yet this comparison is difficult because print advertising with images was rare previous to the Civil War. For reasons that were both technological and cultural, newspapers and magazines carried few advertisements before the last decades of the nineteenth century.[1] Handbills, city directory ads, and individual tradecards were the most common form of advertisement in the mid-nineteenth century. These media are generally categorized as "ephemera" in library archives, and it is exactly this ephemeral quality that makes these early advertising documents rare. There is, however, another way to determine the images used in advertising before the Civil War: by examining collections of "stock" printers' plates. Printers used a limited

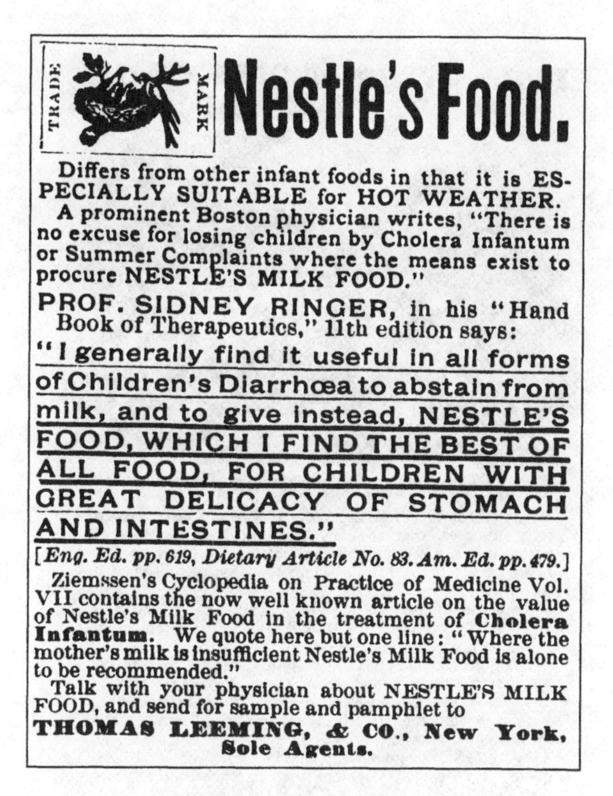

FIG. 5.1. Nestlé's infant food advertisement. From *Ladies Home Journal*, May 1888.

number of plates over and over when a client desired a printed image to accompany an advertisement. For example, a quick look at a mid-nineteenth-century city directory reveals a few images used repeatedly to advertise various merchants and grocers, such as a group of barrels, or a live pig or cow. By comparing these images to the ones found in later magazine advertisements, we can identify some changing trends in the ideas surrounding food provisioning.

One of the earliest and most ornate milk advertisements comes from a Brooklyn City directory of 1840. This promotion for the Jamaica Milk Dairy shows a milkmaid carefully tending her cow (fig. 5.2). The directory ad also states that the cows are not fed distillery waste. While the dairy is right outside the city, the image used to portray the product is pastoral.

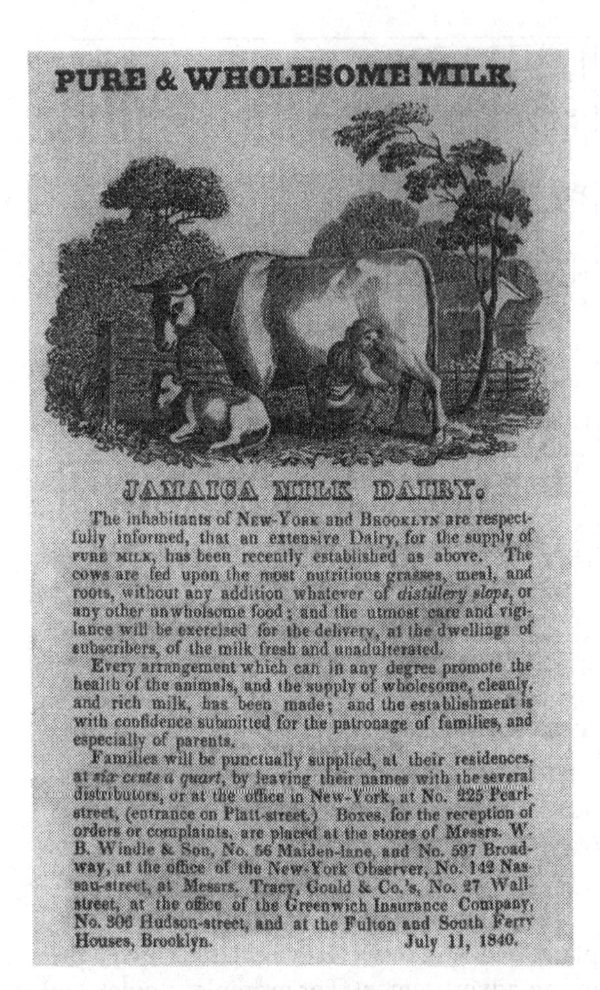

FIG. 5.2. 1840s advertisement from a Brooklyn dairy featuring the milkmaid and cow motif. © Collection of the New-York Historical Society.

Milkmaids and cows fit naturally into the pre–Civil War stock printers' images. Many of these images were taken from popular prints of the time, and the milkmaid and cow was a favorite theme in these engravings and photogravure prints. Food advertisements therefore borrowed from popular art and featured pastoral themes. These images tend to represent food in their country context—live standing animals

FIG. 5.3. Stock advertising cuts of milkmaid images from the antebellum period. From Saxe, *Old Time Advertising Cuts and Typography.*

or fruit on the vine. Yet printers also used these images as decoration beyond the basic goals of promotion. Likewise, the milkmaid engraving used to advertise the Brooklyn dairy is a stock advertising "cut," used on handbills, billheads, and other advertising ephemera. Figure 5.3 features copies of the cow and milkmaid advertising cuts produced by the Boston Type and Stereotype Foundry during the antebellum period, images that were often used to decorate the billheads of dairy merchants. In this case, the milkmaid is a pastoral image that represents purity of the product. The country, the cow, and the caretaker represent the idea of the countryside providing pure, unadulterated produce to the city.

The pastoral idea, of country and city existing in a harmonious relationship, predominated in landscape art throughout the seventeenth and early eighteenth centuries. The pastoral advertising image therefore portrays the goodness of the food through the goodness of the country. City reformers like Robert Hartley worked within this pastoral viewpoint. He envisioned the salvation of infants through the bringing of pure milk from the country. In the same way, he acted for the salvation of orphan children by moving them out of the city and into country homes.

These pastoral images illustrate a continuing agrarian outlook on the nation. They represent an era when cities still looked to the rural areas for their ideas of perfection. Perfectionist reformers like Hartley exemplified this movement of upstate Burned-Over District ideas brought to the early metropolis. Therefore the representations of goodness and purity drew on rural, pastoral images. The pastoral also represented a particular relationship between the city and the country. From the agrarian viewpoint, the farmer was the citizen responsible for the good of the nation, of American society as a whole, by providing both the city's nutritional sustenance and its moral values. This vision of the Jeffersonian idea—including a national mission of pioneer settlement over the frontier during this period—formed the basis of an agrarian politics that put the interest of the farmer foremost on the national agenda.

Pastoral images also represent a particular relationship between humans and nature, a balancing of nature and culture, often mediated by the rural female. The farmer/milkmaid therefore played a major role in the American attitude toward nature, culture, cities, morality, and civilization, and this attitude put rural people at the forefront of the fulfillment of God's will.

Conversely, exposés of the poor quality of "city milk" played on the irony of male milkmaids and city environments, emphasizing the maleness of city milkers, seedy ones at that. Figure 5.4 is from a series of exposés of the city "swill milk" industry published in *Frank Leslie's Illustrated Newspaper*. The caption's reference to "milkmaids" is an ironic comment on the use of country images in claims of purity. The image of the milkmaid, in contrast, depicts a mythical system of milk production

Attack of the "Milkmaids," led by one Stephen Smalley, Upon one of our Artists, in Skillman Street, Brooklyn, Between the Stables. (From Frank Leslie's Illustrated Newspaper, May 15, 1858)

FIG. 5.4. Drawing of swill dairy "milkmaids," *Frank Leslie's Illustrated Newspaper*, May 15, 1858. From New York City Department of Health, *Is Loose Milk a Health Hazard?*

carried out by clean, healthy women caring for their cows and nourishing the cities with clean country milk.

Because of their association with the purity of a product, milkmaids and other pastoral images were used as representations of quality for other products (fig. 5.5).

Carolyn Merchant's book *The Death of Nature* follows the development of views of nature from what she calls "organic" to "mechanical" views a century earlier. The use of the milkmaid image in food advertisements recalls that earlier representation of the countryside. Organic nature, Merchant shows through a survey of literature and art, was associated with the female as nurturer. Of course, as she notes, this "depended on a masculine perception of nature as mother and bride whose primary function was to comfort, nurture, and provide for the well-being of the male."[2] The image of the milkmaid tending the cow and overseeing the purity of the milk for her clients reflects this connection between motherliness and the purity of the natural product. As an extension, this image also represented the nurturing, feminine countryside producing

FIG. 5.5. 1880s trade-card with milkmaid illustration to promote kitchenware. Collection of the author.

food to nourish the active, rational, and therefore masculine city. According to Merchant, ideas about the relationship between country and city followed the Platonist viewpoint during this period. Ideas belonged to the human, masculine world. Therefore, the city, a product of man's ideas, was the imprint of masculine intervention in nature.

The pastoral view of nature, which made rural places feminine and cities masculine, had existed since ancient times and was, even by the time of the first milkmaid advertising images, in decline. Yet "pastoral imagery could easily be incorporated into a mechanized industrialized world as an escape from the frustrations of the marketplace."[3] The milkmaid, overseeing the purity of milk, therefore symbolized this nostalgia for the nurturing countryside, this idea of perfection as the antidote of the downfallen city.

Yet after the Civil War, pastoral images quickly became minor players in food advertising. In milk advertisements, the milkmaid and cow played an increasingly minor role, replaced by pictures of happy, healthy children. Mysteriously, both the human and the animal producers of milk disappeared from dairy product advertisements after the Civil War and, with the exception of Borden's Elsie, were not again prominent in dairy ads until the rise of Ben and Jerry's ice cream.

To a great extent, these changes in the portrayal of milk do not just represent changes in ideas about city-country relationships, they also signify changes in the type of milk advertised. Women's magazines did not contain advertisements for fresh cow's milk. Instead, the most prominent advertisements in these publications were for infant formula and condensed milk, a form of canned preserved milk. These preserved milks increasingly replaced fresh cow's milk as a substitute for breast milk for the infants and young children of the middle and upper classes. The companies in these ads—Nestlé, Carnation, Borden's—and the competitor companies they would soon purchase, such as Anglo-Swiss, represent some of the earliest multinational food companies, producing in both Europe and America. Even more than the fresh milk system (in which they were often also involved), the preserved milk product sector represented a highly industrial form of production, organized along "modern" lines as semi-independent product divisions and with production systems taking advantage of large economies of scale.[4]

Yet images of nature were not entirely absent from the advertisements for these products. Instead of the cow, the most prominent natural image displayed in both Nestlé and Borden's advertisements was,

FIG. 5.6. Nestlé and Borden advertisements featuring birds. From *Ladies Home Journal*, May 1901 (Nestlé's); *Delineator*, April 1914 (Borden's).

surprisingly, the bird, particularly the bird as mother, as shown in the early ad for Nestlé's infant food (fig. 5.1). The bird image fit in well with Nestlé's name, which means "nest," while for Borden's the bird image evolved from its original nationalist image of an American eagle (fig. 5.6). While the bird image changed character greatly over the years, Nestle and Borden's used versions of this animal to portray their products for nearly half a century.

Clearly, advertising images change over time because the new image does a better job at selling the product (and because the product itself changes, as in the change from fresh cow's milk to infant food). But why, it is important to ask, were these newer images so successful? Even infant foods—and "milk modifiers" added to cow's milk to ostensibly make it more like mother's milk—were from cows, which were still milked by hand even if not by women. Nevertheless, babies and birds were for decades the predominant images in these ads. In fact, the baby and the bird were more compatible with the new ideology of consumerism that existed at the end of the nineteenth century.

The analysis of representative images, and how these images change over time, reveals a great deal about the society where these images occur. It tells something about the author of these portrayals, in this case the preserved milk companies, as well as something about the audience for these images, since an image is used frequently only if it is successful with its intended audience. The use of particular images has a politics, which Stuart Hall, in his studies of images of race, has termed the "politics of representation."[5] Feminists such as Merchant have also recently explored the sexual politics of gendered images used to represent various aspects of society. The study of gendered images of a food's source, the countryside, and the food's destination, the body, are both pertinent to the study of milk.

As noted above, Merchant has explored the changes in the representations of the countryside and the removal of women from that representation. Another feminist scholar, Emily Martin, has explored the changes in the representations of the body and the increasingly masculinized representation of bodily existence.[6] Milk is a crucial nexus between these two important sites in the politics of representation of food, since milk is the only food that is produced within the human body, specifically the bodies of women. Looking at the representations of milk, we can see the changes in ideas of nature and motherhood as linked to changes in the organization of the economy, particularly the rise of cities, industry, and governments as the voices of authority in the culture of mass consumption.

The next sections will look at two "sites," or arenas, in which images accomplished the powerful imposition of new forms of authority over the production of American bodies, particularly children's bodies. Those two arenas are the preserved milk advertisements in women's magazines and the milk promotions carried out in American schools. The images used in these promotions represent prominent voices in Americans' changing ideas of perfection at the turn of the century.

NEW VISIONS OF PERFECTION: FROM GOD'S DESIGN TO GOD'S WILL

Darwinism "delivered the *coup de grace*" against ideas of natural theology and pietist religion "by propounding a mechanism of organic change that seemed to eliminate the last trace of wise design in the formation of species."[7] The combined impact of Darwin's ideas and the Civil War had "a serendipitous symmetry. . . . For, just as the shots at Fort Sumter brought the antebellum era to a close, so did Darwin's book."[8] For many theologians, *On the Origin of Species* represented "an assault every bit as deadly as a cannon shot to their reconciliation of science and religion."[9] Darwin had worried that his new ideas about change in nature would have such an effect. In his autobiography Darwin states he was "so anxious to avoid prejudice" that he resisted writing "even the briefest sketch" of his theory of evolution, fearing that it would lead to such a public turn against religion.[10]

According to the theory of evolution, nature no longer represented the result of God's original design; it was no longer static. Instead, Darwin's idea of natural selection described a world created not by God but by natural processes. Other scientists, such as the geologist Lyell, had been increasingly referring to nature in process terms, even before Darwin's publication.[11] Late-nineteenth-century Americans were ready for a more dynamic explanation of their world. In a rapidly changing society, the idea of nature as static could no longer hold. The drastic new dynamic view of nature better represented what Americans of the late nineteenth century saw around them.[12]

Yet most Americans interpreted Darwin through a continuing filter of religious ideas in which God continued to play a role in the lives of people. Discussion of God's role in society and nature changed from one that emphasized the discovery of God's design to one that emphasized history as the triumph of God's will. The historian John Greene traces the modern worldview to this combination of Darwin's ideas of evolution with the Protestant idea of salvation. These new "modernist Christians"

> preferred to regard history as a redemptive process in which man evolved through higher and higher stages of culture, gradually sloughing off habits and attitudes inherited from an animal past. In this view of things, science was a chief instrument of progress. By tracing out the laws, mechanisms, and stages of evolutionary advance, it provided the basis for an evolutionary natural theology. Progress, not the wise adaptation of structure to function, became the proof of divine superintendence.[13]

This modernist Christian view became the major theological perspective on social change from the Civil War to World War I. It was embraced particularly by the new professional class in service to government, such as John R. Commons and Richard Ely, who propounded a "Christian sociology" in the late nineteenth century that would later secularize and become the institutional economics of the early twentieth century. These professionals declared that humanity could not evolve into more civilized beings without the right institutional environment. Professionals therefore needed to design these institutions to allow the correct evolutionary path to take place. Institutions, well designed, would become the vehicles that moved history along the path to progress.[14]

Christian modernism brought salvation to science. God's role was not the original creation of a perfect world. Instead, God guided the world as it changed. His will led the world to a more perfect state. Historical change, then, became the manifestation of God's will, or "providence." Human societies displayed their reverence toward God by working toward what they perceived as God's intent for the future. Yet only history would show what, in fact, God had intended.

The idea of providence predated Darwinism. As the new science ushered in by Newton, Bacon, Galileo, and Kepler brought to the fore more open, changing views of nature, theologians forged a new harmony between science and religion. American Protestants revised Paley's natural theology argument into the doctrine of divine providence, which emphasized God's continuing care and tending of the human race. Lewis Saum, in *The Popular Mood of Pre–Civil War America,* notes that "in popular thought of the pre–civil war period, no theme was more pervasive or philosophically more fundamental than the providential view."[15]

As a result, the idea of perfection changed from the idea of design to the idea of development. The question was no longer "What does God want us to be?" but "What does God want us to become?" A comparison of the differences in Lincoln's two inaugural speeches illustrates this transformation in American thought. In his first inaugural address, Lincoln insisted that he was not about to end slavery, stating, "I have no purpose, directly or indirectly, to interfere with the institution of slavery in the States where it exists."[16] Only dissolution of the Union, he insisted, would lead him to declare war. However, in his second inaugural speech, Lincoln declared that slavery had been a cause of the war after all. "All know that this interest was, somehow, the cause of the war."[17]

Lincoln's change of heart was not simply political rhetoric justifying what had happened. He expresses national surprise at the outcome of the last years, both at the intensity of the war and the emancipation of slaves: "Neither anticipated that the *cause* of the conflict might cease with, or even before, the conflict itself should ceased. Each looked for an easier triumph, and a result less fundamental and astounding." Lincoln was trying to explain why things had turned out differently than he, or the country, had intended.

His answer was to attribute the great, wrenching changes brought about by the Civil War to God's will. He notes that both the South and the North believed that God was on their side, that they each were the beneficiaries of God's providence:

> Both read the same Bible, and pray to the same God; and each invokes His aid against the other. It may seem strange that any men should dare to ask a just God's assistance in wringing their bread from the sweat of other men's faces; but let us judge not that we be not judged. The prayers of both could not be answered; that of neither has been answered fully. The Almighty has his own purposes. "Woe unto the world because of offenses! For it must needs be that offenses come; but woe to that man by whom the offense cometh!" If we shall suppose that American Slavery is one of those offenses which, in the providence of God, must needs come, but which, having continued through His appointed time, He now wills to remove, and that He gives to both North and South, this terrible war, as the woe due to those by whom the offense came, shall we discern therein any departure from those divine attributes which the believers in a living God always ascribe to Him?

For Lincoln, the Civil War was a test of two different human perceptions of God's will. Neither perception was entirely right. The war brought God's retribution for Americans' sin: not realizing God's will. Even the North, through the horrors of the war, was being punished. Yet the North, the winning side, represented God's true will. God's providence, his will, reveals itself not in the initial design but in the final outcome. The theologically important idea of Revelation becomes the revelation of God's will in history. Social action thereby becomes a testing—trial and error—process by which humans with different ideas about God's will act upon their beliefs. The best result comes from discovering and following God's plan. In this way, humans perfect society by seeking change that they believe to be in accordance with God's goals for the future.

From this transformation of God's role in the world came the story of progress, and the industrial path taken by American society. Industrialization—in part due to the war itself—was one of the vast changes brought about by the war, and was proceeding at a rapid pace. As we have seen, the progress perspective links God's will to the industrialization and technical development of society.

This progress perspective also brought a new envisioning of the downfall story. From the downfall perspective, the speedy trip of urban and economic growth was a Faustian agreement, a movement away from God. The negative effects of industrial growth—pollution, poverty, monopoly power, bad health—were signs that providence had abandoned human society because it had taken this path. From the progress perspective, alleviating the problems of industrialization required further effort, but the path was essentially a good one.

FROM PRODUCER TO CONSUMER

By the turn of the century, a new consumerist ideology became the most prominent expression of the progress idea of perfection. American food advertising's images of perfection changed from production to consumption, just as American politics moved from the predominance of agrarian to urban interests. In milk advertising, the image moved from the productive milkmaid and cow to the consumer images of the child and the bird.

In the case of milk, at least in its preserved form, the progress notion of perfection was most prominently represented by the beautiful child. Advertisement after advertisement, from the Victorian tradecards that predated magazine advertisements to prize child advertisements of the 1950s, featured the child's body as a primary sales image.

The Victorian obsession with idealized images of children reflected this new perspective on perfection. In a society where change was occurring rapidly, and often seemingly outside human intervention, the happy, healthy child became the focus of control. Whether selling milk, drugs, or dress, advertisers pushed their products as the way to gain the perfect child. The "prize child" became the ultimate perfect product. Possessing beautiful, healthy children was a powerful indication that a family had the blessings of divine providence.

In milk advertising, for either fresh or condensed fluid milk or for milk "modifiers" and "lactated foods," these new consumer images of perfection replaced earlier pastoral images that had represented the care taken in the product's production. For the advertiser, moving the locus of attention from the site of production to the product of consumption was important, especially in those many cases when the product was in fact deficient in needed nutrients.[18] In many cases, testimony by mothers that the product was responsible for the child's survival and health accompanied the child's picture. In one Eskay's Food advertisement, the mother testifies that she has fed her son this food since his first month of life and that "it has built him up a perfect boy, bright, healthy and never one sick day."

In this new view of perfection, the nature involved in the production of the product also changed. Rather than the nurturing feminine countryside, nature became "matter," a fuel used for the sustenance and growth of human machines, particularly children. While, according to Merchant, this mechanical view entered scientific thought centuries earlier, the emphasis on the body as a human machine entered the popular imagination through patent medicine advertisements around the time of the Civil War, which were so successful that they were later extended to food advertisements.[19] In milk advertisements, this resulted in a drastic change in the images portrayed to sell milk: they began to emphasize the product of milk input—the healthy, happy child.

This prize child, pictured in endless milk advertisements, represented the product of good mothering. The chubby bodies, the obviously upper-class names, and the smiling faces no doubt prompted mothers to compare their own children to these products of lactated milk foods and condensed milks. Advertisements encouraged such comparisons with comments such as the number of pounds a child weighed at a certain age (one even boasting an astronomical twenty-eight pounds at ten months!), and the stories of how the child was saved from the door of death because it would eat "nothing else" (fig. 5.7).

Interest in perfecting children peaked with the formation of the "Better Babies" movement. Generally carried out by local women's clubs or even the formula makers themselves, the movement generally

FIG. 5.7. Ladies' magazine advertisements featuring idealized children, 1901–1925. From *Ladies Home Journal*, June 1901: *Delineator*, May 1915 and June 1925.

involved prize baby contests, in which children were judged according to a scorecard, the winner generally being "the plumpest, the fairest and the most blue-eyed," much like the children in milk advertisements.[20] The movement encouraged women to compare their children to a general standard of physical beauty. Because formula, and especially condensed milk, makes children gain weight more quickly, breast-fed children appeared increasingly below standard. By the 1930s, this voice of authority had become overpowering. Advertisements were encouraging mothers to make comparisons between the perfect "test children" in advertisements and their own (fig. 5.7).

Merchant argues that the mechanistic view of nature was male-centered.[21] The triumph of this view of nature also had implications for the representation of agriculture and rural places. The city, rather than gaining succor and nourishment from the countryside, sought to improve and manage the countryside to meet urban needs. The industrial vision of perfect farming, to be discussed in chapter 7, involved just such a plan by urban people to improve the quality of rural life and to make agriculture more efficient to better feed the cities.

The other most prominent image used to advertise milk was a bird. Why a bird? Answering this question once again uncovers the complex nature of relationships (of people to each other, particularly within the family) and embodiment (the relationship of people to their own natural bodies and—particularly for women—the bodies of their family) in the industrialization of food. Once again, looking at the ways these issues were represented in food advertisements can give us some clue about how advertisers attempted to intervene in these relationships.

An overview of early-twentieth-century food advertisements shows that the image of the "maker" of the food was often problematic. Instead, a "replacement" maker or provider of food was often portrayed. The Quaker in Quaker Oats was one of the first successful portrayals of a replacement maker that consumers identified with. In many cases racial or childlike "others" (someone who could not be confused with a consumer's actual local baker or grocer), such as the black chef used to advertise Cream of Wheat for over a century, represented the "maker."

In milk formula advertisements, an obvious replacement for the maker would be the cow. However, instead, the bird became the prominent symbol of milk formula. The examples in fig. 5.6 are not exceptional; they are standard advertisements for infant formulas in the early twentieth century. Borden's advertised its "Eagle Brand" condensed milk through a fatherly bird-authority figure. Nestlé's advertisements evolved from the tiny image of a bird feeding its children in a nest (fig 5.1) to a dominating image of the protective stork. Why did these companies feel the need to replace the cow with the bird?

To explain this phenomenon, we must look not at the food itself, but at the ideal process that the advertiser wants to encourage. A bird gathers food and brings it home to the family nest. A cow produces food within its own body. In other words, the cow too closely represented the process of embodied feeding, while the bird more accurately represented the mother participating in a process of commoditized consumption, that is, shopping and bringing the infant's sustenance back home. The "replacement" nature icon, in this case, becomes part of a discourse about embodiment, nature, and family relationships around milk.

THE RISE OF NUTRITION IDEOLOGY

In the mid-nineteenth century, the milkmaid represented the producer as an authority in control of the purity of the product. With the rise in science and rational control over nature and the concurrent rise in urban control over the countryside, authority was no longer embodied in the pastoral image of the milkmaid but in the industrial image of the expert. This new "overseer" of purity, the inspector and the veterinary doctor, was always depicted as a man. Therefore, advertisements replaced the nurturing milkmaid in her generative partnership with nature with the expert male who made pure milk through scientific inspection of the subject, the cow. These public health images therefore inevitably show the veterinarian holding his stethoscope up to the animal's lungs, a monitoring, not a motherly, gesture (fig. 5.8).

These changes in worldview also reflect changes in demography and the consequent political changes that went with them. By the 1920s, urban dwellers had gained a majority over the rural population and politicians and businesses had to take into account these changes in their constituencies. The transformation of the countryside from mother to managed could not have taken place without these demographic changes.

The otherness of nature was therefore overcome through management. Not only was the cow domesticated, the bacteria in her milk was now under control. Nature was perfectible through human intervention, and milk now led the way as "the perfect food." Women, no longer in charge of nature, were instead grateful recipients of nature managed by industry and government. Women and children were themselves managed through nutrition campaigns that predominantly took place in schools.

One of the most prominent of these efforts was the school health campaign. In 1915, dairy industry interests organized the National Dairy Council to research and promote milk and dairy products.[22] Its promotion efforts have been "aimed primarily at school children."[23] In addition, various states set up local milk/dairy councils and boards to promote consumption. State boards were often funded by assessments on milk sales, generally voluntary until the 1970s.[24]

In 1919, just as the end of World War I was threatening to spill a surplus of war-supplied milk onto the market, the U.S. Department of Agriculture and the dairy councils started a cooperative program to

FIG. 5.8. Picture from a public health publication showing a veterinarian holding a stethoscope to a cow. From New York City Department of Health, *Is Loose Milk a Hazard?*

promote milk drinking in the schools. USDA specialists had found "a relatively high percentage of undernourishment among children and that this condition frequently accompanies a low average per capita consumption of milk."[25] In response, the USDA created a plan to promote milk drinking through intensive, weeklong local "milk-for-health" campaigns, through local agricultural extension agents. The campaigns were based on the premise that the "failure to use an abundant supply of milk can not be attributed wholly to a lack of material wealth, but rather to a lack of information regarding the importance of milk in the normal development of the growing child."[26] Weight and height were the major measuring sticks used to determine the health of a child. The milk-for-health campaigns promoted weigh-ins as the determination of child health. According to Pirtle's history of dairying,

> The educational milk campaigns in which the government and the National Dairy Council cooperate are generally undertaken to reduce undernourishment in children and to improve the general health by increasing the consumption of milk. These campaigns last from one to two weeks during which time the value of milk is stressed in every way possible. A later survey is made to ascertain the development resulting from the campaign and the increase in milk consumption. Many cities have used these campaigns to bring the use of milk up to a proper standard and in general when such a campaign was properly handled the results were much greater than expected. This is one of the principal causes for the great increase in milk consumption in the United States. In 1917 the per capita consumption was 42.4 gallons, and in 1920 it was 43 while in 1925 it was 54.75 gallons.[27]

The campaigns specifically targeted schools and included speeches by home economists, the distribution of publications, and special activities such as poster contests and the performance of milk "plays" and milk songs. Informational materials, as well as milk plays and songs, were generally supplied by the National Dairy Council. Children were weighed before and after the campaign. The USDA claimed that "an average of 12 per cent reduction in undernourishment among school children has been accomplished in these communities following the milk campaigns, while the increase in milk consumption ranges from 10 to 30 per cent."[28] Aubyn Chinn's 1924 published curriculum, titled *Health Habits: Suggestions for Developing Them in School Children*, represents a typical example of campaign material. A nutritionist "in Charge of School Health Programs" at the Philadelphia Inter-State Dairy Council, Chinn stated that "[t]hese lessons are the outcome of health work conducted for a year in schools of varying types of children, ranging in age from six to twelve years."[29] The purpose of the book was to "establish the daily practice of those habits that make for health and strength in such a manner that the child takes pleasure in their performance."[30]

The book contained various lessons, stories, games, and play scripts for children to perform, each covering particular good health practices. Included with the lesson book were various milk education pamphlets published by various national and state dairy councils, including a beautifully illustrated book and poster published by the New England Dairy and Food Council. These publications featured a "Good Health" fairy, who advised children to develop eight health habits, from daily toothbrushing to adequate amounts of sleep. The advice included drinking four glasses of milk a day.

This practice continues today. For example, the Hartford Public Schools Department of Food Services and Nutrition Education announced in its "Nutrition News" Web page (vol. 6, Issue 6, 1998–99) that a group of Hartford schools participated in a breakfast milk promotion "which includes 'Got Milk' gifts during the months of April and May."[31]

While the promotions have changed over the years, the 1924 lesson plan contains many of the points that have been typical of these promotions through many decades. A boy character representing protein builds muscle and "lives" in meat, although "[t]oo much meat is not good for girls and boys."[32] Fannie Fat lives in butter (and not margarine) and is an important component of the diet. Minerals and vitamins live in fruits and vegetables, "But suppose we can't get all of these things every day?" The teacher is instructed to present the milk bottle as containing everything in these other foods. It is the complete food that will protect children and their development, even if they don't eat enough of the other good foods.[33]

Nearly every American child can remember a milk promotion campaign carried out in his or her school. One of my students remembers coming home after a milk promotion program to announce to her parents that ice cream was, in fact, good for you, and then demanding some. Earlier generations of children were handed literature such as the National Dairy Council pamphlet *Milk Made the Difference*. This promotion derives from a University of Wisconsin research project as published in the university's cooperative extension bulletin. This bulletin graphically showed the results of experiments feeding rats vegetable oil or butterfat. Pictures show a scrawny, undeveloped, oil-eating rat next to a plump, prosperous-looking, butter-eating rat. The writers wasted no time in comparing the results of these experiments to what could happen to human children.[34] Part of Chinn's curriculum is a dairy council pamphlet featuring various animals either fed milk or not fed milk after weaning.[35] The results were graphically illustrated in the difference in size between the dogs, chickens, and pigs who received milk and those who were denied it. The pamphlet claims that the animals were fed the identical ration, except that only one of the pair received milk. "Faulty bone growth is the inevitable result of a ration which does not include mineral matter and the valuable vitamines [*sic*] which are found in milk."[36] In this case, the implication is that milk is the only place to find the nutrition necessary to grow. The pamphlets do not consider the idea that other foods might have included these nutrients and could have been provided in the ration.

Newer studies have challenged earlier nutritionists' research and their results. For example, the food historian Celia Petty reexamined an earlier milk feeding experiment on children inspired by the animal feeding studies described in the National Dairy Council pamphlets. Her reexamination showed that the spectacular growth gains shown in these studies actually recorded "'catch-up growth' in a group of chronically malnourished children [the studies were carried out in an orphan asylum] and not 'super-growth' in a group of already adequately nourished children, as it was claimed."[37] The foundation for this misinterpretation of the scientific results, Petty claims, was ideological, since the results of this study "were used to support the view that much of the ill health from which the poor suffered to a disproportionate degree could be attributed to exclusively dietary causes. Poverty was thus reduced to a 'deficiency' disease." Yet the dairy council publications did not simply advocate more milk drinking, they provided an entire ideology of the healthy body, especially the child's body. Dairy council publications, because they became general child health publications used in schools, had a strong voice in defining the healthy child's body and behavior in America. For example, the New England Dairy and Food Council children's book and poster, with the "Good Health" fairy, does not simply advocate more milk drinking. The milk message is embedded in a larger message about health, including bathing, exercise, more fruit and vegetable consumption, and adequate sleep.

Clearly, these are good recommendations, and the dairy councils' role in promoting good health habits in children is in many ways to be commended. Promoting good health habits is a worthwhile goal that public schools often find necessary as a way to improve the health of their school population and, hopefully, the American population. Yet nutrition education is a politically sensitive issue: Who gets to define "good health" and "good health habits"? Who gets to define the perfect body?

Legislation in the 1980s made funding through assessments on milk sales mandatory for all dairy producers. This funding went toward national campaigns such as "Milk: America's Health Kick" and "Milk: It Does a Body Good" in the 1980s and the milk mustache and "Got Milk?" ad campaigns today. Most dairy promotion has focused specifically on increasing fluid milk and, to a lesser extent, cheese sales.[38]

However, it could also be argued that milk promotions may have led to health problems. The campaigns began to promote more and more milk per day to more and more groups of people. According to the nutritional historian Harvey Levenstein, "In 1926 experts were recommending that children up to the age of eighteen should drink one pint of it a day. By 1937 this was the recommended intake for adults; the under-eighteens were up to one quart a day."[39] Levenstein notes that the expenditure of ten cents a quart for twenty-eight quarts a week for a family of five "would normally have taken up about half the food budget of a relief family of that size in 1936.[40] Yet both the USDA and the private dairy councils were making this recommendation at the time. Milk promotion campaigns may therefore have influenced poor families to spend more money on milk, leaving less money for fruits and vegetables, that after the first glass or two of milk, may have had a better food value for the dollar.

Did the promoters of this large increase in milk drinking have evidence that a quart of milk a day was the right amount for a child to drink? The most often used evidence to support their claim was the increase in weight of children who drank more milk, along with the ubiquitous pictures of the milk-drinking versus the non–milk-drinking baby animals. Yet, according to Levenstein, nutritionists could not prove the link between dietary deficiencies and inadequate milk drinking, or even that health in general was seriously deficient in the country as a whole, even during the Great Depression. "The repeated failure of the dreaded health problems to arrive did prompt some questioning of [nutritionists'] lugubrious assessment of the nation's nutritional state."[41] Infant mortality rates and deaths from vitamin deficiency diseases continued to decline during the Depression decade.[42]

Yet few researchers have ever questioned the continuing close relationship between the milk councils and school health programs. In an era when state governments sue tobacco companies for the public health costs of cigarette promotion, governments might want to be especially cautious about close relationships between themselves and a product that has been connected with a major American health problem: heart disease. Few people have questioned the use of private industry information to promote health in our schools. Are dairy councils really disinterested parties in school public health programs?

MILK PROMOTION AS ENLIGHTENMENT

The public blindness to the problems of letting an interested party dictate food habits to the American public has had to do with the perfection story's dependence on "enlightenment" as a way of achieving common goals. Enlightenment as a component of the perfection story is based on the idea that perfection, once discovered, must be taught to the uninitiated. The economists Spencer and Norton celebrate the fact that "[v]arious demonstrations and other means of publicity have convinced a large proportion of the people of the value of milk as a food."[43]

Why do public authorities rely on health information provided by an interest that could not be neutral? Because no one questioned the perfection of milk and its role in perfecting children's bodies. The groups promoting milk are presumed to be working in the general interest. Schools that are hard pressed to afford informational material on children's health have therefore been happy to accept the pamphlets, the plays of yesterday and today's teen "milk ambassadors" as well as other promotions offered by the dairy councils. Because milk is an unqualified good, its promoters are above suspicion.

Only if we begin to question the perfection of milk can we see the extent to which the dairy councils have gained authority over the definition of children's health. Only then can we begin to question why we allow an organization with private interests in promoting the consumption of a particular product to be a major voice of authority on health in our schools.

Medical evidence now tells us that the milk promotions of the past were not an unqualified good. A large proportion of the American population—who were brought up on school "weigh-ins" and weight gain contests—are now considered seriously overweight. Heart disease is our number one health problem, in part due to America's overconsumption of animal fats, particularly dairy fats. These statements shouldn't be taken as a condemnation of milk as a food. The problem is that the government allowed a private organization to define as healthy a quart of milk a day, an amount that was ridiculously high, without any evidence that this was necessary. This is particularly problematic considering the fact that all milk at the time was whole milk—the high-fat kind. Yet the quart-a-day recommendation went unquestioned for decades. The image of milk's perfection provided a set of intellectual blinders.

These blinders hid an inconvenient fact: that the promotion of milk was in the economic interest of dairy farmers. As noted above, the government began to collaborate with the national and local dairy councils in the promotion of milk at a time when a large milk surplus threatened due to the end of World War I. As chapter 7 will show, with the rise of the fluid milk industry, a dairy farmer's income increasingly depended on the amount of milk going to fluid markets. If consumers are convinced to drink more milk, dairy farmers' income rises. For this reason, many dairy farmers, and their governmental representatives, saw the increased consumption of fresh milk as a solution to dairy farm income problems.

In other words, the promotion of milk in schools was not simply a child health program, it was a farmer income program. Yet because milk was "perfect," public health officers did not see a problem with this connection. If they did, they considered it simply a win-win situation for both consumers and farmers. As vendors of the perfect food, dairy councils became the authority on the perfect body.

Another way to think about it is to ask the question, "Why not cheese?" Dairy councils promoted butter to convince the public away from its major competitor, oleomargarine. Fluid milk promotion enhanced dairy incomes. Cheese had no non-dairy competitor, and increased cheese consumption did not do much to improve dairy farmers' incomes. As chapter 7 will explain, cheese became a surplus control item, a way to get milk out of the fluid market, generally by dumping it into the manufacturing sector at a low price. The only way to increase the price a farmer received was to increase the amount that went to fluid milk markets.

In some ways, the lack of cheese promotion was a good health decision on the part of dairy councils. Cheese during most of the twentieth century was full-fat. Ironically, "skim cheese," the type recommended today, was the term used to define bad-quality cheese, which had some cream skimmed out before manufacture, sometimes "filled" with other kinds of fat. Farmers in this case were succumbing to the temptation to make both butter and cheese from the same milk.

However, dairy councils failed to promote cheese not because of concerns over American fat consumption. From the early-twentieth-century view of nutrition, cheese was as much a "complete" food as milk. It contained the same basic nutrients in a more digestible form. But a rise in cheese consumption did not raise farmers' incomes to the same extent, so it received much less attention from government or dairy council promotions.

Needless to say, many of the officials and experts who advised wives and mothers to increase their family's milk consumption were part of educational and state institutions in dairy industry states. No one would bother to raise rats on butterfat in Wyoming or Texas land grant universities as they did at the University of Wisconsin. In many land grant institutions, nutritionists advising increased milk consumption worked virtually across the hall from agricultural economists trying to control milk surpluses. Obviously, increased consumption seemed like the ultimate solution. Therefore, reinforcing the "indispensability" of milk in the mind of the public has been carried on relentlessly until only recently, when studies found that there may be such a thing as eating too many dairy products.

Why would one question the goal of improving bodies through the consumption of milk? Aren't lowering the death rate and improving children's bodies worthwhile goals? Didn't the rise of the milk system contribute to these goals?

The contributions of dairy products to human health are unmistakable. Food historians have noted the relationship between dairy diets and higher statures.[44] Over and above the contribution of dairy products to health in particular farming areas throughout history, dairy has played a significant role in providing cheap protein and other nutrients to city populations. As with earlier farm populations, it is important to note that much of this dairy protein was not in fresh fluid form, particularly in other countries, like Britain. Cheese provided significant amounts of much-needed protein and fat to the British working-class diet. This was particularly true before the 1880s, when meat became more available to this group.[45]

Has milk promotion improved the health of Americans? The important role of milk in the American diet, as it turns out, does not have to do with the perfection of milk but its use to correct the deficiencies caused by other American lifestyle practices. For example, the well-known American nutritionist Jane Brody disputed the claims of the Physicians Committee for Responsible Medicine, countering that the research simply does not prove that milk (in its nonfat state) causes cancer, heart disease, or osteoporosis. She does, however, admit that

> It is true that in most Asian countries, where little or no dairy products are consumed, there is a much lower incidence of osteoporosis than in the United States. But it is also true that Asians eat a lot more calcium-rich vegetables and a lot less protein than Americans do. The excess protein consumed by most Americans actually removes calcium from the body. Asians also get a lot more physical exercise and consume less cola, which can impede the use of calcium.[46]

In other words, Americans need to drink milk to make up for other, less healthy aspects of their diet and lifestyle.

Dairy products, to some extent including fluid milk, also played a significant role in supplying certain European working-class and rural populations of the late nineteenth and early twentieth centuries with essential proteins and fats not otherwise available in their diets. While Americans of all classes ate enormous amounts of meat in the nineteenth and early twentieth centuries (and comparatively few fruits and vegetables), the average British diet tended heavily toward bread and was relatively protein-scarce, particularly since meat in a British family was usually reserved for the male breadwinner.[47] Ireland is another example where milk played an important nutritional role: the milk and butter added to their potato-centered diet provided the Irish with their only source of calcium and vitamin A, plus some protein and fat.[48] While Americans now eat less meat than their predecessors, the diet of the late nineteenth and early twentieth centuries was not a scramble for essential proteins, fats, or vitamins as it was in Britain, since the meat diet provided most of these. Despite constant harangues by nutritionists for the poor to substitute milk for meat, the poorer classes that could not afford sufficient meat protein generally could not afford milk either.

However, it is clear that certain regional American diets of the nineteenth century were deficient in a number of nutrients that milk could have supplied. The addition of more milk to a diet low in protein and B vitamins certainly helped to lower the incidence of pellagra (a B vitamin deficiency disease) in the South.[49] The addition of milk to the diet could have helped prevent certain nutritional "wasting" diseases. However, milk reformers at the time could not recommend milk for its ability to help prevent these diseases because the relationship between these diseases and specific nutritional deficiencies was not yet clear. In addition, nutritional deficiencies had many causes beyond the lack of milk. Interestingly, the existence and nutritional importance of vitamins were not discovered until 1912, when researchers, searching for an advantage of butter over margarine, discovered the presence of vitamin A in butterfat.[50]

MILK, RACE, AND NATION

The idea of God's will manifesting itself in social success also meant that the most socially successful were the visible manifestation of God's providence. This combination of Protestant theology and Darwinism, basically social Darwinism, manifested itself in ideas about social perfection. The dominance of the white race in both local social and world orders became one more sign of God's providence, a sign of racial superiority. Because milk was a food of northern white Europeans, the link was soon made between this food and white social dominance. By World War I, American rhetoric associated milk with the food of an imperial nation and superior race. A National Dairy Council publication distributed in the 1920s quotes the famous nutritionist E. V. McCollum:

> The people who have achieved, who have become large, strong, vigorous people, who have reduced their infant mortality, who have the best trades in the world, who have an appreciation for art, literature and music, who are progressive in science and every activity of the human intellect are the people who have used liberal amounts of milk and its products.[51]

McCollum's ideas were not unusual. Ulysses Hedrick's history of agriculture in New York state, published in 1933 (interestingly enough, a milk strike year in the state), made a similar point:

> A casual look at the races of people seems to show that those using much milk are the strongest physically and mentally, and the most enduring of the peoples of the world. Of all races, the Aryans seem to

> have been the heaviest drinkers of milk and the greatest users of butter and cheese, a fact that may in part account for the quick and high development of this division of human beings.[52]

These statements were intricately tied into nationalistic sentiments. McCollum's "newer nutrition" was credited with building and maintaining soldiers' health during World War I. An alarming number of British conscripts, when examined for service, were found to be malnourished. The solution at the time involved importing huge quantities of condensed milk from the United States (a purchase that truly expanded that industry) to feed to soldiers. Subsequent examination showed rapid diminishing of malnutrition. The building of the national body became the national goal in the postwar era.

National Dairy Council pamphlets often represented the racial superiority of milk drinking more subtly, through pictures. In Figure 5.9, "Elizabeth K" provides an example of an undernourished girl. She is notably dressed in an old-fashioned, "Old World"–looking outfit. Her dress and hair are arranged in a way that would suggest to the reader of the time an immigrant child. "Virginia," on the other hand, is a milk drinker, and milk drinking has not only made her more the exemplar of "health and beauty," it has also made her more of a modern, American child, with her sailor dress and pageboy haircut.

Social workers in their visits to immigrant families encouraged the notion that eating American food was the path to health. They discouraged the eating of spicy ethnic foods and nonmeat dishes such as spaghetti.[53] Therefore, milk became not only one of the reasons for Northern European white racial superiority but also a way to pass that superiority on to other races and ethnicities.

CONCLUSION

The human body is the nature at the end of the food production chain (as the nature of soil and sun is at the beginning of the chain). Ideas about how consumers relate to their own bodily natures are just as important as how they conceive the nature around them. The idea of milk as a perfect food emerged along with the idea of the perfectibility of the physical body, particularly the child's body.

Elizabeth K
One of many undernourished high school girls.

The High School Student

IS your daughter starting High School this fall as thin, tired and undernourished as this young girl, Elizabeth K? Her strained, over-serious face, hollow chest, stooping shoulders, flabby muscles and evident "posture of fatigue" tell a story of lowered vitality and overwrought nerves, which make her totally unequal to the heavy demands her school work will make upon her. It is a crime to tell such children to "stand up straight," "hold your shoulders back," "put your chest out," because their muscles are absolutely unequal to maintaining such positions, and it is also true that a large number of undernourished children are found to have actual spinal curvature in greater or lesser degree.

Virginia
This is the way a school child *should* look, ready for work and play, and proving that *Health* is *Beauty*.

FIG. 5.9. Illustration from 1921 National Dairy Council publication, *What Milk Will Do for Your Child*. © Courtesy of the Historical Collections at the New York Academy of Medicine Library.

As this analysis shows, the relationship between consumer consciousness and the rise of industrial food is more complex than current narratives about the industrialization of food currently acknowledge. The idea of perfection changed significantly with the advent of Darwinism. The perfect society was no longer the result of discovering God's design. Yet the ideas of perfection and God's role did not disappear. Instead, in the Progressive perspective, perfecting society became the discovery of God's will—or the perfect future, often expressed in a more secular way. This idea of perfection was well suited to a society in the midst of rapid change and one that was becoming increasingly consumer-oriented.

Private and governmental milk promotions reflected these changes. Milk's purity and quality were previously represented by the pastoral source of production—the cow and milkmaid. This was replaced by the prime icon of development during this period: the child's body. The cult of the child epitomized this new consumerist idea of perfection, the perfect child being the testimony of the discovery of God's will in the transformation of nature. From religious authority to industrial and governmental authority, male authorities mediated the relationship between women, their families, and their bodies.

Yet the irony of this analysis is that the discourse separating female nature from milk did not originate with the industrial voice. The consumer voice, a part played in this discourse by the male urban reformer, had already reorganized the relationship between body, country, and food. In fact, it took the industry itself time to catch up with this ideology, abandoning the milkmaid as the representation of purity and replacing it with the child, the bird, and the veterinarian.

Only with the rise of the city comes the need for a government with the capacity to arrange for collective services such as sewers, water systems, transportation, and the organization of utilities into "franchises" so as to make an investment in electrical lines profitable. Needless to say, city growth led to a greater need for an expanding city government and an expanding set of powers for city officials. Therefore, putting the veterinarian instead of the milkmaid next to the cow to represent milk's purity was not just the creation of a new myth but a reflection of the rise of a new form of state authority.

Milk "education" campaigns hammered home the benefits of milk to the point where the food became considered an essential in any diet. The problematic side effect of this Great Milk Propaganda Campaign is that consumers began to demand a reliable and inexpensive milk supply all year long. Milk became so indispensable to American society that possible shortages brought on cries to regulate the milk trade in the same ways that were being considered at the time for the electric or railroad industries, which were characterized by "an obligation to provide an adequate supply at all times."[54] Some authors even suggested that milk be regulated as a utility.[55]

If milk was so necessary for good nutrition, it needed to be available at a low price all the time, even at those times of year when cows did not normally produce much milk. Therefore, while early consumer groups were primarily concerned with milk safety issues, later groups were more concerned with milk prices, which they claimed were artificially high. Consumer groups also had the advantage of unity. Unlike farmers and handlers, who were fractured along both spatial and economic lines, consumer groups had no reason for division in their demand for cheap, safe milk. This certainly helped this group achieve its interests. It is also obvious in the politics of the day, particularly the urban politics, that urban milk drinkers represented an enormous number of voters. Assuring a safe and inexpensive milk supply was certainly a major plus to any politician interested in remaining in office, as Mayor Fiorello LaGuardia realized when he set up a system of poor "milk stations" in New York City. The provision of cheap, safe milk was clearly a quick way for politicians to work their way into the hearts of their urban constituencies. This urban constituency and their desire for cheap, safe milk would provide the prevailing form of milk politics in New York state, a form of politics with substantial repercussions on the state's dairy farmers.

no [illegible] consumer groups had no interest in [illegible] in their demand for cheap [illegible] community between [illegible] interests. It is also obvious that the politics of [illegible]. Particularly, the
urban groups that can make demands to extend an enormous num-
ber of voters. Assuming a safe and inexpensive [illegible] supply, we can
[illegible] a [illegible] policy [illegible] community in [illegible]
[illegible] system of poor
[illegible] in New York City [illegible]
[illegible]
their [illegible] and for
cheap [illegible] form of [illegible] politics
[illegible]
state [illegible]

PART II

PRODUCTION

6

Perfect Farming

The Industrial Vision of Dairying

IN 1903 THE Rockefeller Foundation carried out a study of dairy sanitation with the New York City Department of Health. The result was a book called *Clean Milk.*[1] This book laid out a system for the production of milk free of harmful bacterial contamination. The illustrated frontispiece foretold the nature of the system described within. The picture is of a man in a clean white uniform, his head covered with a clean white cap, milking cows in a pristine environment (fig 6.1).

In contrast, Milton Rosenau's *Milk Question* presents a graphic illustration of the undesirable "cheap labor" that needed to be eliminated from milk production (fig. 6.2). Unlike the white-capped milker, the ragged man posing with a pitchfork outside a barn exemplifies all that city people feared about the state of their country milk system. City people expected their milk to be produced by "bacteriologists," and the pictured farm laborer did not meet the potential for such a professional transformation. The picture's caption is ambiguous: Is this man a farm laborer or the farmer himself?

The "clean milk" picture was meant to illustrate the dairy of the future. The white-capped milker appeared in many public health publications concerned with dairy sanitation at the time. The uniformed man bending carefully under the cow is not meant to represent the farmer. One cannot imagine farmers waking up in the morning and donning such garb. The man is clearly an employee, and the farmer is entirely absent from the picture.

As noted earlier, there was one other person commonly presented with cows in public health publications at the time: the veterinarian, also in a white coat. The veterinarian's gesture is one of control: he is monitoring the health of the cow through his stethoscope.

To make their point, public health publications relied heavily on pictures: pictures of unsanitary barns and employees compared to the new barns, equipment, and high-producing cows organized for sanitation and efficiency. Yet in all these pictures, the farmer is invisible. While city health officials admitted that "it is to the co-operation of the dealers and farmers that we must look for an absolutely pure milk supply," farmers are hardly ever pictured in public health material.[2]

With the demise of the milkmaid image, the image of the producer disappears entirely from the consumer discourse on milk safety. As we saw in chapter 4, the farmer was also largely missing from discussions about the milk question in New York City. In page after page of public health publications over the decades, cows stand either alone or with the white-uniformed employee or the white-coated veterinarian. The dairy employee and veterinarian most accurately represented the urban vision of a sanitized countryside reformed to meet urban needs. The veterinarian monitored the cow and provided the expertise to keep her running smoothly. The employee provided labor. In this vision of dairying, the farmer as "manager" was merely the conduit between the expert and the employee.

FIG. 6.1. Frontispiece, Belcher, *Clean Milk*.

IT IS IMPOSSIBLE TO MAKE GOOD MILK WITH CHEAP LABOR AND CHEAP METHODS. IT IS IMPOSSIBLE TO MAKE BACTERIOLOGISTS OF SUCH MEN

FIG. 6.2. Illustration of a "nonscientific" farmer/farmworker. From Rosenau, *The Milk Question*.

The pictures of cows and their non-farmer tenders in public health publications were meant to convey the supervision of cows by modern people, the doctors or the wage earners who do the hand-work in a modern organization. For example, most of the thirteen "prominent dairymen" pictured in Pirtle's history are not farmers. Dressed in professional attire, they are industrialists, farm journalists, and scientists at dairy colleges. Pictures of modern cows and professional men are meant to convey progress, and farmers do not generally look like modern, professional people, even when they are.

Why did the story of milk as a perfect food erase dairy farmers from the picture? To answer this question, we must ask what ideas the pictures in public health publications were meant to convey, and how these ideas differ from those conveyed by the earlier milkmaid image. In both cases, the person associated with the cow represents labor, yet there is a fundamental difference in the meaning of "labor" in these two pictures. Anson Rabinbach argues that the social meaning of labor changed in the late nineteenth century, with Helmholtz's discovery of the first law of thermodynamics.[3] Helmholtz's law established the universality of energy as the single force in the universe. It was Marx who first applied this idea to labor, in his concept of "labor power." "Labor power" was the energy necessary to make nature's materials into human products, and Marx showed, following Helmholtz, that the nature of labor power was simply a force of energy that was the same from one worker to the next. Interestingly, Marx's concept of labor was taken up by neoclassical economists in their analyses of labor as a "factor" of production. The difference between Marx and the neoclassical economists was the addition of other forces of energy to the analysis of economic value, namely, capital and land. While Marx calculated value entirely according to the expenditure of labor power, neoclassical economists conceived of production as the mixture of three separate factors: labor, capital, and land. The neoclassical revolution in economics that occurred at the turn of the twentieth century introduced the concept of marginal cost. Production from a marginal cost perspective reduced farming to the three universal factors—land, labor, and capital—which the farmer, with expert advice, managed with the goal of increasing productive efficiency.

Rabinbach argues that this concept of labor differed radically from earlier conceptions, exemplified by Locke in his explanation of property.[4] Locke and others in his time conceived of labor as something unique to an individual, a person's creative ingenuity and skill, which, "mixed" with nature, produced that which an individual could claim as property, under natural law. Rabinbach refers to this as a "metabolic" view of labor, and the image of the milkmaid represents this earlier conceptualization. The milkmaid represented someone who was a caretaker, a tender of nature, not simply a source of energy. Yet neither is she a bacteriologist, making pure milk by applying science to the process. The white-capped milker is a standardized being producing a standardized service, as organized by science. As Marx noted, the force the worker exerted was no different from that provided by a machine. In fact, most such workers would be replaced by milking machines as dairy farms were electrified in the 1930s and 1940s. The milkmaid represented purity through an interactive relationship to the cow. The worker, on the other hand, represented sanitation through the scientific control of his labor, as monitored by the scientist.

Stories—and their accompanying images—are powerful tools that take hold of society in different ways. As part 1 of this book showed, the perfect story of milk played a strong role in forming the way American society has engaged the body and the self. The second part of this book will focus on public discussions of dairy farming, dairy farmers' social movements, dairy regulation, and the current debates over biotechnology and organic food. This examination will show that the perfect story has played a powerful but complicated role that explains not just why Americans drink milk, but how they make it as well. The industrial bargain between farmers, consumers, and large-scale processors, discussed in part 1, created seemingly inevitable and inexorable effects that transformed dairy farming.

The clean milk images represent a vision of perfection for dairy farming as portrayed by a rising class of urban health and agricultural professionals. Like the discussions about safe milk, the discussions about good farming were controlled increasingly by professional groups. The "Country Life Movement" of the Progressive Era exemplified the influence this new group of "amateur and professional social scientists" had on the future course of rural society. "Prescriptive, naïve, self-confident, and brash, social scientists did not doubt that the public interest could best be served by social and governmental institutions which modified and controlled society according to scientific principles." While some Country Lifers idealized rural life, others in this group "were struck by the rudimentary nature of rural social and political institutions. Not only did social scientists believe that rural institutions were too weak to respond to rural needs, but they also feared that their weakness allowed many country individuals and communities to isolate themselves from institutional influences."[5] The key to solving these problems was to make the rural economy more efficient.

Not surprising, New York's agricultural officials, particularly Cornell agriculturalist Liberty Hyde Bailey, were dedicated Country Lifers. New York's agricultural establishment embraced the vision of an efficient, business-like dairy sector. Their vision coalesced with that of public health officials in many ways. Public health officials concentrated on the cleanliness of the operation, while agricultural economists and extension agents focused on the efficiency of milk production. Both agreed that the large, technologically sophisticated, capital-intensive operation, with employees milking cows and owners managing the operation using the advice of science, would best serve the farmer and the consuming public.

In figure 4.2, Rosenau portrays the difference between the two systems in his comparison of the menacing small business dealer system and the rationalized, efficient modern system of milk provision.

The perfect story was therefore an urban story, a product of the industrial bargain between consumers, farmers, and large milk firms. The city, in its vision of agriculture, reduced the countryside the same way the capitalist reduced the worker—from a social being to a mere provider of force, either cheap labor or cheap food. While farmers collaborated in the industrial bargain, the industrial vision of production ended up remaking them in ways they could never have predicted.

As with the perfect story of consumption, the perfect story of production requires that the elements of the story become the universalized, standardized, and under the control of those telling the story—in this case, the experts and their urban constituency. The perfection of production therefore required universalization and standardization the three inputs, or "factors" of production: labor, capital, and land. Each factor was a homogeneous input added together to create goods. Inputs varied according to quality, but this quality was judged on a single standard: the potential to produce. Factors—such as crop land, cows, employees—were less or more productive, and no other characteristic of these factors mattered.

The agricultural college in the dairy state with the strongest urban constituency—the College of Agriculture at Cornell University—was the most forceful representative of this industrial vision of dairying. We will examine this industrial vision, focusing primarily on Cornell's agricultural bulletins and reports, to show that neoclassical ideas about the productivity of labor, capital, and land made the intensive market milk dairy farmer more "perfect" than other dairy farmers. The industrial vision focused on transforming dairy agriculture into the management of the three inputs or "factors." In the industrial vision of dairying, farmers managed labor, capital, and land to create the most efficient and clean form of production.

THE INDUSTRIAL VISION OF PERFECT LABOR

Cornell economists began their yearly surveys of dairy farmers in the early 1900s and have continued the surveys ever since.[6] Cornell started these surveys in response to controversies over milk prices at the turn of the century. The New York state legislature requested the surveys to pinpoint whether or not farmers were responsible for a steep rise in milk prices. Interestingly, the surveys showed that farm prices were not high enough to provide sufficient income for the average dairy farmer.

Yet, over time, these surveys began to serve another purpose: they were used to show that larger, more intensive farms produced better income for farmers than smaller farms. As a result, what began as a way to provide data about consumer price problems became an indicator of farm problems, which then became a tool to resolve those problems by transforming dairy agriculture.

The surveys measured labor efficiency in "man-work units," which represented "the average or normal amount of work accomplished by one man in one day."[7] The management surveys showed that the most efficient farms, analyzed in terms of these units of efficiency, were the ones that produced the most milk per man-work unit, and that these farms tended to be organized along industrial lines. For example, the surveys showed that in good years, "[f]or any sized farm a relatively large herd of cows was advisable,"[8] mostly because of greater production efficiency: "the major advantages of a large farm business were found to be in labor efficiency, capital efficiency, and horse and power efficiency, and in having larger receipts."[9]

To become more efficient, the farmer needed to become more technically educated as well, since "[t]he handicap which the untrained man must overcome is constantly increasing."[10] Bulletin after bulletin related the education of the farmer to the size of the farm, since "[t]he operation of the law of advantage and size in successful farm management often hinges upon managerial ability and financial resources of the farmer, as well as the ratio of price outlays to price returns."[11] As sociologists of agriculture have noted, the factor that distinguishes agriculture from other forms of production—and that keeps it from becoming fully industrialized—is seasonality. The seasonality of work on a farm—peaking at planting and harvest—tends

Table 6.1
Seasonal Variation in Milk Production, United States, 1930–1981

Year	% of Monthly Average Peak	% of Monthly Average Low	Low Month as % of Peak Month
1930	127	83	65
1940	127	81	64
1950	127	80	63
1960	120	87	73
1970	113	91	81
1980	107	95	89

Source: Manchester, *The Public Role in the Dairy Economy*, 50.

to make farms organized around family labor rather than wage labor more competitive, at least in certain commodities.[12] The cow's lactational cycle tends toward spring "freshening" (milk after pregnancy and calving), making it a seasonal production system.

Yet with the rise of the fresh milk system and the urban focus on milk as a way to build children's bodies, consumers began to demand fresh milk all year long. In response, market milk farmers changed the cow's production cycle so that a certain number of cows lactated in the winter, thereby evening out milk production throughout the year. Because market milk dairying overcame dairying's seasonality problem, it held the promise of an industrial form of labor organization based on year-round hired employees.[13] Cornell farm management economists noted that "farms with high output per man also had more uniform seasonal milk production than did those with low output,"[14] although U.S. dairy farms, on average, produced only two-thirds as much milk in the winter as in the summer until the 1960s (table 6.1).

Cornell agricultural economists were very much aware of the diversity in dairy farming in their state. They published their farm management surveys by region, showing the differences in farming from one part of the state to the next. Many summary reports also distinguished variations in dairy farming by region, often characterizing regions broadly according to the map in figure 6.3. These studies showed that farmers responded to differences in natural resources and market access by instituting different strategies on the farm and participating in different economic networks beyond the farm gate.

After World War II, however, studies increasingly concentrated on the benefits of intensified, specialized, dairying. While 1930s studies noted the benefit of including cash crops, such as cabbage and beans, in the dairy operation, later bulletins noted that, "farms with high output of milk per man placed more emphasis on milk production and less on other enterprises."[15] Yet the farmer who became specialized in dairying in fact became diversified in terms of the kinds of activities needed to maintain the cow. Cows managed intensively had to eat special high-protein forages and less pasture. The industrial cow needed to be housed and fed, and the farmer had to manage the manure. Therefore, the specialization in dairying was, for farmers, a taking on of the work previously done by the cow. The cow no longer provided the work of her own self-management. Industrial farmers began to take on a dual role as both cropper and cow-tender. In areas where dairying had been part of a diversified system, this transformation usually meant the elimination of other crops on the farm in favor of growing feed for the cows.

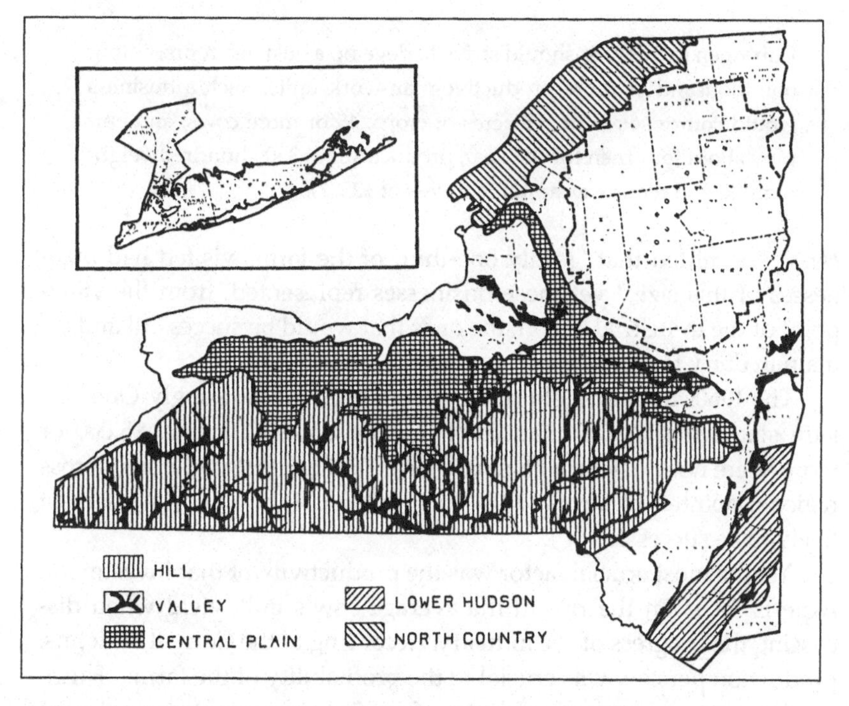

FIG. 6.3. Map from Cornell agricultural bulletin showing dairy farming regions. From Cunningham, *Commercial Dairy Farming in New York.*

CAPITAL: THE INDUSTRIAL COW AND HER APPARATUS

The industrial story of dairying usually starts with the cow. Centering the history of dairy farming around the cow enables the modernist story to be told: of a biomachine that was becoming more and more productive. The industrial vision of perfect farming envisioned large dairies with high-producing cows.

As mentioned above, increasing herd size—the capital investment in animals—was a critical factor in achieving higher farm incomes. In the 1930s, Cornell farm management surveys published data results by region, in consideration of the differences in availability of productive inputs from one region to the next. Nevertheless, the advice on herd size was similar for bulletins across regions: all 1930s regional farm survey bulletins tended to point to the farms that were larger than average as more likely to make a good income. For example, in the farm management survey bulletin for the Tully-Homer area in 1936, E. G. Misner advised,

> Dairymen in this area should strive to develop a business representing one thousand or more productive-man-work units. Such a business would comprise about 115 acres of crops, 45 or more cows, and employ about four men for the year, produce about 3500 hundredweight of milk, and require a capital in excess of $25,000.[16]

Misner admitted that "[o]nly one-third of the farms visited had businesses of this size," yet these businesses represented, from the viewpoint of the industrial vision, the ones that would be successful and remain in dairy farming.[17]

The 1960s bulletins told the size story even more strongly. One bulletin stated that "[t]he chances of making a good income, $5,000 or more, were nonexistent with small herds."[18] The 1960s bulletins, across regions, pointed to farms with "greater than 50 or 60 cows" as most likely to be successful.[19]

Yet the most crucial factor was the productivity of that investment. Experts focus on the rise in the average cow's milk yield when discussing the progress of the industry. According to the Cornell bulletins, production per cow was crucial to the profitability of the farm: "Farmers who obtained rates of production from 25 to 50 percent above the average in each area did well financially," stated one of many bulletins on the subject.[20] The cow has been the primary biomechanical investment that, with scientific "improvement," raised dairy farm productivity.

Histories of dairying therefore laud the rising average milk yield per cow. The tripling in per-cow milk production over the last few decades is, in fact, a remarkable achievement. This achievement is the result of the work of decades of cow-breeding and feeding research, especially the development of high protein feeds and artificial insemination. Progress histories rightfully celebrate rising milk yields as a major achievement of modern science.

The endless pictures of cows in Pirtle's *History of Dairying* were meant to convey the desirability of the new, modern purebred animal that gave more milk.[21] Dairy histories often talk about, and even individually name, "star" cows that produced larger than average amounts of milk. Pirtle's history described cow breeds and particularly high-producing "champion" cows, lauding "[t]he progress of the purebred cattle in America."[22]

Pirtle's history, based in the industrial vision of dairying, emphasizes the poor quality of cows before the rise of scientific breeding. He describes the original American cow as a wretched, ill-kept, badly bred animal. In fact, many contemporary writers made the same point. According to one particularly critical contemporary commentator, "Either the cows are the most worthless breed, or they are, as is most generally the case, utterly unprovided with nourishing winter food."[23] These commentators also stated that farmers did not know how to feed cows, how to choose animals that produced adequate milk, or whether the cows were profitable or "boarders" (did not pay for their keep). According to this story, farmers treated cows badly, and did not breed cows for production, because they did not know how:

> Farmer neglect of the dairy clearly did not result from commercial unimportance but from inferior knowledge of breeding and feeding techniques, their preferred concentration on land-intensive food-crop production, and poorly developed markets. Over the remainder of the century, as farmers' knowledge improved so did dairy yields.[24]

In the perfect production story improved management, improved knowledge, and growing markets go hand in hand with the creation of agricultural progress. Profitable markets lured farmers into enlightened production, with the help of experts. Enlightened farmers purchased, bred, and fed higher-quality cows. Experts and promoters of the industry exhorted farmers to pay more attention to breeding, eventually through artificial insemination, a practice vigorously promoted by the agricultural journalist and editor W. D. Hoard.[25] Hoard is pictured as one of the "prominent dairymen" in Pirtle's book.

As with the perfect story of consumption, the perfect production story depends on the idea of enlightenment, and the history of agriculture is described as a series of revolutions in which farmers awoke to the possibilities of efficient production. The historians Louis Schmidt and Earle Ross provide a typical description of this process of change in farmers' attitude that "broadened the farmer's outlook and awakened him to a realization of his educational needs and opportunities," so that each successive period

> witnessed the rise of a new generation of farmers who were ready to abandon the old methods of farming and adopt new ones, once their utility had been demonstrated. Agriculture thus liberated from the fetters of custom and tradition, was prepared to enter upon a new era of scientific development.[26]

The "awakening" metaphor, at the base of the idea of the Enlightenment, contains within itself a curious form of politics. It implies that no internal process has taken place, that change comes not from endogenous social contexts and based on particular social, historical, and environmental circumstances, but from exogenous forces. These external forces tend to be the information and innovations provided by modern science and neoclassical economics, findings that are "diffused" among the unenlightened from centers of learning.[27] Centralized information creates the perfect vision and remakes agriculture in its own image of homogeneity and efficiency.

A branch of rural sociology research—generally referred to as "adoption-diffusion studies"—has looked at the rate of technology adoption and why some farmers do or do not adopt a technology. Working from the neoclassical perspective of modern farming as the perfect ideal, these researchers label nonadopters "laggards" and blame their lack of technology use on individual personality traits, a form of unawakened-ness.[28]

By the 1960s, substantial mechanization was under way on dairy farms in New York and many other dairy states, and the greater efficiency of the dairy apparatus on larger farms was added to the list of its advantages. Accordingly, 1960s bulletins stated that larger farms tended to be more mechanized.[29]

Market milk farming therefore functioned according to an industrial logic: maintaining income required increasing production, which required increasing inputs of capital and labor. Much of this increase in production involved a decrease in seasonality of production; farmers fed and milked cows for a larger number of months a year. Therefore, dairy farmers who milked during the winter needed to grow, buy, and store significantly more grain and hay. This usually required an extra investment in production machinery and storage facilities. It also required more acreage of higher-quality land for feed production.

The modern industrial cow is a biological factory for the transformation of feed into milk. The farmer increasingly became the tender for this increasingly expensive machine. As tender, the dairy farmer had two functions: to "fuel" the cow and to prevent breakdown. In the cow's case, breakdown means disease. Dairy farmers continually talk about how far they can push their cows without a bout of illness.[30]

To increase milk yields, market milk farms began to use a "hotter" (more protein-enriched) feed. This required farmers to change their cropping habits, spending more time and effort growing high-protein crops that would fuel these high milk–yield animals. Farmers purchased increasing amounts of feed "concentrate" and cultivated more acres of alfalfa, corn silage, and corn for feed. In response to the need for high-protein feed, dairy scientists innovated new feed crops and feed storage systems. These innovations included the silo, corn silage, and other high-protein forages such as alfalfa.

One of these intensification strategies, the increase in alfalfa cultivation, can be seen in census figures over the entire period under study. Alfalfa is a protein-rich forage that increases milk yield and can be dried for winter feeding. Historically, it has required high-quality land resources to grow productively. Alfalfa acres per cow therefore represent another measure that characterizes the farming practices of intensive dairy farms.

While alfalfa strains more acclimated to poor soil conditions were later developed, there was a particularly acute relationship between land resources and the ability to grow alfalfa in New York in the 1930s:

> The alfalfa region of the State is closely associated with the limestone regions. Alfalfa requires a deep, well-drained soil containing lime, and this is provided in the Mohawk Valley and the Finger Lakes areas. The hill areas of the Southern Tier counties are not suited to alfalfa production mainly because a large proportion of the soils are thin, acid, or poorly drained. In the St. Lawrence lowlands most of the soils contain lime, but are too shallow or poorly drained for alfalfa.[31]

Early varieties were also more sensitive to cold. Although, as one farmer once noted to the author, agricultural experiment station breeders have not yet developed an alfalfa that "grows in a swamp," newer varieties are less sensitive to moisture and temperature. As a result, more alfalfa has been grown in areas with these poorer conditions.[32]

Alfalfa is a relatively recent addition to dairy farm feeding strategy, as shown by the growth of acres of alfalfa per cow in the United States (table 6.2). The rise in alfalfa acreage was part of the "package" of farm practices in market milk dairying. In a study of farms in Livingston County, Warren noted the relationship between acres of alfalfa and an increase in work units per acre, which "makes possible a larger business with the same acreage."[33] An industrial farm, therefore, used fewer acres per cow but used each acre more intensively through the cultivation of hotter feeds.

Other strategies of feed intensification include increased cultivation and feeding of corn and hay silage, increased growing of corn for feed, and the purchase of feed "concentrate," all of which provided the extra boost of protein so important to increasing milk yields. Acres of ensilated corn per cow have increased along with alfalfa. Areas with poorer soils are more likely to grow corn or hay silage rather than alfalfa. However, most New York farmers do not grow corn for concentrate feeding, since it requires better soil resources than are generally available for this purpose. Therefore, farmers generally purchase most corn and other concentrate elements. As with alfalfa, intensive dairy farms associated with fluid milk markets have tended to grow more of these high-protein forages and to purchase more concentrate.[34]

As the list of city milk sanitation regulations grew longer, the investment and resources necessary to make market milk increased substantially. The need for extra sanitation measures—including tested cows, barns with cement floors, and milk cooling apparatuses—made

Table 6.2
Increase in Acres of Alfalfa Planted,
United States, 1927–1986

Year	Area Planted (thousands of acres)
1900	6
1927	11,401
1930	11,565
1940	13,908
1950	17,970
1960	27,654
1970	27,202
1980	26,269
1986	26,748

Source: DuPuis, "The Land of Milk."

fluid milk production into a more capital-intensive operation. Warren described this change.

> There was a time when the requirements for producing fluid milk were little more than the requirements for producing butter, that any milk that would run was ready on any day to be added to the fluid milk supply. Under such conditions, the price of fluid milk was determined by the price of by-products. It is now essential to recognize that the production of fluid milk is an entirely different industry from the production of butter and cheese. It is an industry that has to assume an obligation to provide an adequate supply at all times.[35]

Market milk farms therefore bore greater risks, since they made higher investments, carried larger debt loads, and pushed a biological milk production system (the cow) to greater production that also increased risks of disease. Incomes tended to be higher, although they also tended to be continuously under assault because the price of this "necessary" food was a politically sensitive issue, as will be described in chapter 8. As described in chapter 4, each part of the market milk system tried to impose the extra costs of making milk safe onto the other. Consumers and dealers did not want to pay the amount necessary to make safe milk, and farmers did not want to "foot the bill." The extent to which farmers *did* in fact foot the bill also entailed taking on more risk.

LAND: PROFITS DEPENDENT ON GOOD SOILS

No matter what the environmental and economic context, Cornell applied the "work unit" standard to its practices. These agricultural economists found that the addition of an extra work unit in areas with high-quality resources resulted in a higher income than that extra work unit applied in a lower-resource area. Therefore, while resource and market diversity was recognized in these studies, dairy farms were primarily judged according to a single standard: the potential for further intensification, as measured by response to additional labor. The bulletins illustrated these differences on graphs such as the one in figure 6.4.

Bulletins noted that achieving an adequate farm income required "plenty of good land."[36] In general, farms with good soil resources tended to be situated in the often narrow river valleys that striped through most of the state. These "valley" farms could produce significant amounts of alfalfa and corn silage. In areas farther from milk markets, farms on good soil tended to be diversified "general" farms growing nondairy crops as well as tending cows.

Dairy farms on poor soils were generally located on the hilly slopes rising above these river valleys. These "hill" farms comprised over 30 percent of dairy production in 1944, but they were much less likely to practice the intensive, industrial farming envisioned by Cornell economists.[37] As the next chapter will describe, these farms were involved in a significantly different type of dairy farming.

In bulletin after bulletin, Cornell economists recorded the inability of hill farms to raise productivity and thereby produce an adequate income. By the 1960s, good valley soil had become critical to the farming success story. For example, one bulletin stated that the potential of a high income was "nearly twice as good on a valley farm as on a hill farm." A 1961 bulletin noted that 25 percent of the hill farmers did not make a positive labor income, because "some farmers are just not able to make the adjustments that are necessary in changing agriculture."[38]

Beginning in the 1920s, Cornell University began a project that eventually mapped every dairy farm into the state's agricultural landscape. "Land utilization" maps linked farms to the soil potential, road access, and other characteristics that land experts used to determine which farms were "permanent" and which were "marginal," that is, unable to provide adequate income and therefore likely to be abandoned.

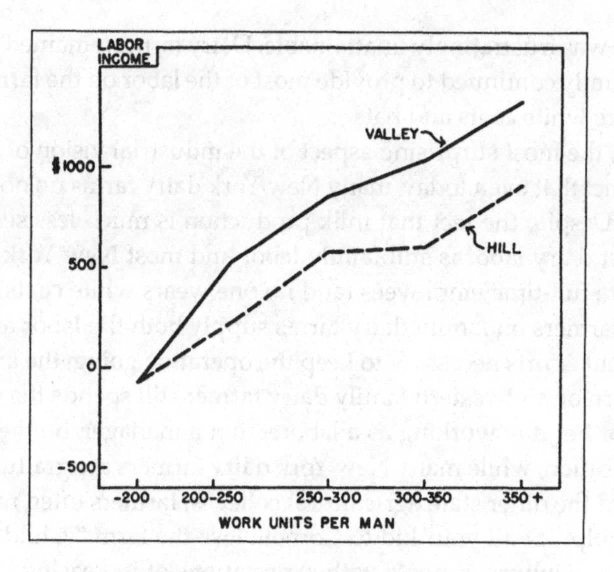

FIG. 6.4. Graph showing the difference in income response to an increase in labor, by valley and hill regions. From Cunningham, *Commercial Dairy Farming in New York.*

Yet the separation of farms into permanent and marginal became a proxy for the distinction of dairy farms into those with the resources to be intensified and industrialized, and those that required the use of more extensive strategies. Extensive hill farms were no longer a different way of farming, they were a less perfect form of production. Environmental officials, such as Gifford Pinchot, were also concerned about soil erosion and advocated reforestation of "marginal" farmland as a state conservation policy.[39]

CONCLUSION

In summary, according to Cornell economists, the perfect dairy farm had a larger herd of cows, good valley cropland, high milk yields, substantial amounts of machinery (including a milking parlor and a bulk tank), and a manager farmer with trained employees. Yet the "perfect farm," whether defined by the white uniform and cleanliness of the public health board or the efficiency and size of the land grant economist, was frustratingly unattainable. Dairy farms remained small, the farm family continued to provide most of the labor on the farm, and no one wore white coats and hats.

In fact, the most surprising aspect of the industrial vision of dairying is the fact that even today, many New York dairy farms do not meet this ideal. Despite the fact that milk production is much less seasonal today, most dairy labor is still family labor and most New York farms do not have full-time employees (and no one wears white coats).[40] Because the farmers on family dairy farms supply both the labor and the management efforts necessary to keep the operation going, the average northeastern or midwestern family dairy farmer still spends the majority of his or her day working as a laborer, not a manager, on the farm. As for education, while many New York dairy farmers are graduates of Cornell and the other state agricultural colleges, farmers often repeat a common quip: "Send your kid to Cornell, lose the farm." Children return from agricultural schools with expectations of increasing the size and intensity of the farm operation. Their parents often see this as a risky prospect, since it requires taking on more debt.[41]

In many ways the industrialization of dairy farming in the Northeast and Midwest has been painfully slow. For example, half of all New York dairy farms in 1992 had fewer than fifty cows.[42] A survey of New York dairy farms in the 1980s showed that 80 percent of farms still did not have modern milking parlors. Most farms continued to use the "pipeline" milk system introduced in the 1950s. Even more telling was the survey finding that a third of all the farmers still used the basic mobile milking machine first introduced around 1910. On these farms, the milk was either poured from a bucket into a "dumping station" or *carried by hand* to the bulk tank.[43]

Most northeastern and midwestern farms do have the bulk tank storage required for tank truck transfer of milk. Midwestern farms are even smaller and less technologically sophisticated. Bulk tank adoption, because it speeds pickup for milk processors, was encouraged by premiums.[44] Yet, as late as the 1970s, 10 percent of all Wisconsin dairy farmers still delivered their milk to manufacturers in cans, much like the farmers of a century before.[45] Except under California's unique environmental conditions, the fully industrial dairy farm has not yet become a reality.

Adam Smith's idea of the "invisible hand" linked the perfection of markets to the perfection of society. According to Smith, capitalists, while only intending to make money, actually promote the greatest good, since the way to make money is to sell the goods that people need most. In Smith's words, the capitalist "intends only his gain." Yet he is "led by an invisible hand to promote an end which is not his intention."[46] Historical change, then, is motored by this hand, which motivates the owners of the production system to make more profit by providing more people with what they need. The market sets the conditions for these changes, including a need to increase the division of labor, to increase the size of enterprises, and to increase the use of machinery to replace labor. These are the major characteristics of modern, industrialized production. Therefore, the invisible hand of the market is the motor that propels the vehicle along the road to progress, defined as industrial production. But the failure to fully industrialize dairying brings us to a necessary question: Is there really such a thing as a "perfect" market?

7

The Less Perfect Story

Diversity and Farming Strategies

LOOKING AT CONSUMPTION, part 1 has shown that milk drinking is as much a product of cultural ideas as it is of material needs. It became the perfect food for the creation of perfect bodies. By questioning the idea of perfection, especially the representation of certain forms of food and certain bodies as perfect—that is, universal and complete—we found that the consumption of milk was a creation of social and political relationships. It is only natural, then, when we turn to the rise of the fluid milk production system, to question concepts of perfection here as well.

To tell a less perfect story of dairy production, we must take into account the diversity of dairy practice in the United States. The question of diversity in the sociology of agriculture is a long-standing one. For more than a century, social theorists concerned with agriculture have tried to answer the agrarian question: Why do small agricultural enterprises—either peasants in Europe and the Third World or family farmers in the United States—still exist? While some agriculture does resemble the factory assembly line, as a whole food production has remained widely diverse in scale and structure.[1]

Rather than simply trying to explain why agriculture hasn't industrialized, sociologists have recently become more interested in explaining the many different ways farmers practice the art of food production. These sociologists are paying particular attention to agriculture in marginal and low-resource regions in order to discover the "adaptive strategies" by which farmers in these areas produce food.[2] Rather than considering small and marginal farmers a remnant of a static past, sociologists of agriculture are realizing that less industrialized farmers practice a dynamic set of adaptive strategies that are embedded in strong social networks. These networks include both economic influences, such as the neighborhood processing plant, and social influences, such as ethnicity, religion, or family relationships.[3]

Sociologists dealing with farmers in the less industrialized world have been particularly interested in the wide variety of ways a single food can be produced. Accordingly, in working with local farmers, these sociologists often focus on understanding the specific "farming systems" prevalent in a region.[4] Once again, the tendency is to focus on local institutional contexts, from markets to farm households.[5] In addition, researchers interested in "agroecology" or sustainable agriculture have adapted farming systems research methods to understand how farmers can produce food without depleting local resources.[6]

In addition, a number of U.S. sociologists have reexamined agriculture in the light of the new "turns" in economic, historical, and political sociology, each of which emphasizes the diversity of practice and the social embeddedness of particular economies.[7] Some of these sociologists have focused on dairy farming, rediscovering regional differences in economic structure.[8] Other sociologists are paying particular attention to the social and institutional networks in which actors in the food system—from farmers to consumers—practice. This approach emphasizes the "polyvalence" of these networks and their potential as "alternative organizational patterns of production and consumption."[9] A number of studies taking this "actor-network approach" emphasize the social construction of food systems and the interlinkages of production and consumption.[10]

In Europe, due to an immediately pressing set of concerns, sociologists have turned to the study of agricultural diversity. Europeans have noticed that the European Union's "Common Agricultural Policy" has threatened to homogenize farming practices across member nations. European citizens treasure their regionally and nationally distinct ways of using the land and producing food. As a result, the decline in regional differences in farming practice is widely viewed in these countries as a tremendous social loss. In response, the European Union is now attempting to create agricultural policies that will retain some of the diversity of regional practice.[11]

To explore more closely the historical relationship between institutional context and strategies in dairy farming requires data that capture the diversity of dairy strategies over time and link these strategies to alternate institutional networks—particularly alternate market institutions—beyond the farm gate. Data linking particular farms to particular market networks would, of course, be ideal. As noted earlier, the Cornell management surveys in fact recorded a great deal of diversity among dairy farms and their institutional networks, even while projecting diversity's disappearance. However, historical and geographic information linking dairy farm strategies and networks is not widely available. The census of agriculture does not ask these sorts of questions. In addition, the census reports data only down to the county level, which is not regionally specific enough to distinguish between different regional dairy practices.

Nevertheless, there is available a "special census tabulation," requested by Cornell, of New York farms by township in 1940 and 1964. Because these data report dairy farm characteristics at the township level, it is possible to combine these data with other local information to get at least some idea of regional differences in dairy strategy. The following discussion will combine these data with two other available forms of information that give some indication of the natural and institutional resource linkages available in dairy townships. First, information on the town addresses of dairy plants during these periods makes it possible to identify each township's major dairy market type—its economic network—and then to examine whether or not farming styles varied according to market type. While it is not possible to directly compare farming strategies of particular market milk and cheese factory farmers, the local nature of markets during this period enables the use of townships as an indirect way to measure differences by market. The following discussion will focus specifically on market milk and cheese factory markets as two widely different economic networks.

Second, soil maps provide information on soil resources generally available to a farmer in a township. While soils vary significantly within townships, for the purposes of this study soils in each township were examined to determine average soil quality, and each township was labeled "poor," "fair," or "good" according to its average soil quality. The soil indicators link farm strategy to soil resources. These town-level farming-market-soil linked data on dairying therefore provide a unique opportunity to examine the linkages between New York dairy farm strategies and their particular economic networks and natural resources, over the span of a quarter century.[12] County-level data from the 1980s will illustrate the final step from diversified dairy farm strategies and market linkages to an intensified, specialized system linked to a single market network.

COMPARING CHEESE AND MARKET MILK FARMING PRACTICES

Perhaps more than other forms of agriculture, dairy production strategies "on the ground" have varied greatly from farm to farm and region to region. In some cases, the reasons for this have to do with farm family choices. To begin with, herd size is often related to particular family circumstances—fewer or more children at various ages, the involvement of siblings or parents in the farm. Often, a dairy farm family will add cows when a son or daughter becomes interested in farming. A larger herd often becomes a way to support two families.[13] In other words, dairy farmers today increase the size of their herd for other reasons besides a desire to be "modern." Yet choices of production strategy, even those that seem like private family choices, are embedded in particular social contexts and networks or economic institutions.

This analysis will use three indicators to determine the prevalent dairy farm strategy in each township:

1. *"Average cows per farm"* indicates the size of the farm. As noted earlier, intensive farms tend to have larger herds. Townships with larger average herd size would therefore have, on average, more intensive farms.
2. *"Average acres per cow"* indicates the intensity of land use. Farm townships with fewer average farm acres per cow had more intensive farms, on average, than those townships where farms were using, on average, more farm acres per cow.
3. *"Acres of alfalfa"* provides an indicator of the intensity of production in terms of feeding. As mentioned in chapter 6, alfalfa is a high-protein feed that increases milk production.[14]

Table 7.1 shows the dairy farm strategies in townships that have either predominantly market milk or cheese factory market linkages in 1940. According to all the indicators, farmers in dairy townships linked primarily to fluid milk markets at this time were more intensive in terms of their herd size, land use, and cow-feeding strategies. Farms in these market milk townships also used land more intensively, as shown by figures indicating fewer acres of land use per cow. These farms also grew (and presumably fed) more acres of alfalfa per cow. Farms in market milk areas also tended to have better soil resources.

Table 7.1
Dairy Strategy by Market Network,
New York Dairy Townships, 1940

Strategy	Market Network	
	Fluid Milk	*Cheese Factory*
Average Herd Size (cows per farm)	21	15
Average Intensity of Land Use (farm acres per cow)	9.6	12.3
Average Feed Intensity (alfalfa acres per cow)	0.28	0.09
Resource Endowments (percent of market type in each soil quality)		
Poor	7	22
Fair	40	60
Good	53	18

Source: DuPuis, "The Land of Milk."
Note: Dairy townships are defined as townships with more than five hundred cows.

On the other hand, farmers in townships associated with cheese factory markets in 1940 practiced a more extensive dairy farming strategy. Extensive farms had fewer cows per farm but more acres per cow. Extensive farmers also tended to grow much less alfalfa, indicating a less intensive cow-feeding strategy. More farm acres per cow and fewer acres of alfalfa per cow indicate a feeding strategy that emphasized pasture over the feeding of the high-protein forage necessary for higher milk production per cow.

Unlike the intensive market milk system, cheese and butter manufacturing networks did not require a change in the cow's seasonal lactation cycle. Following the traditional British "open pasture" system, farmers organized their practice around milking during the time of spring and summer pasture, rather than in winter, when cows needed to be fed more expensive feed rations.[15] Farms that pursued this strategy were generally hill farms, also known as "summer dairies." In contrast, fluid milk dairies, generally valley farms, were called "winter dairies" because they produced milk both in summer and in winter.

Cows fed on pasture rather than feed rations were not record producers, but they did not cost a great deal to maintain, either. Pasture-based summer dairying required significantly less capital or labor investment than winter dairying, since farmers grew little feed and let the cows harvest the hay. Under the pasture system, the cow fed herself easily and cheaply when summer pastures were green. She found her own food in the pasture grasses, and she mostly took care of her own needs. She even fertilized her own food production, by dropping manure in the pasture. She contributed labor to the system. Also, since cows eat less when they are not producing milk, feed requirements to "board" cows over the winter were low and could be accomplished with relatively little labor or capital investment (to store extra forages). In addition, many summer dairy farmers had diversified farm operations, producing fruit, wood, or other products, or supplying labor to local rural industries.

As a result, cows in these cheese factory townships gave less milk, but they were also less costly to feed. In addition, farms in cheese factory townships were more likely to be located in areas with poorer soil resources. These were the areas where a pasture-based farming strategy made more sense. These farms didn't have the soil resources to grow high-protein forages (both alfalfa and corn silage). As a result, farms in these areas tended to depend more on grass, which did not lead to the highest amount of milk production per cow, but which also did not require the intensive use of land.

As these data show, there was at this time a strong relationship between market networks and particular dairy farming strategies. Farmers' resource strategies were heavily linked to the networks they participated in. Farmers in areas where cheese factories predominated used significantly different farming strategies than farmers in areas that primarily sold to fluid milk markets.

Cornell economists were aware of these differences in farming strategies. For example, in a 1931 farm management study, Catherwood noted that "the amount of concentrate [high-protein feed] used in the cheese-factory area is about half of that used in the market-milk areas."[16] He also noted that concentrate fed to cows on farms in cheese factory market areas tended to consist of less "hot" homegrown grains such as oats and barley.[17] While specific studies of cheese factory farming were rare, a 1925 study that did focus on this form of dairying showed that farmers freshened 95 percent of the cows in this area in the spring, since "[t]he success of a dairy farm with this market depends on efficiency in the use of hay and pasture. The herd should be milking when the cheese factories open, and yet the cows should not freshen too far in advance of this date" since there was no market for milk produced by these cows once the cheese factories closed.[18] Yet the bulletin also noted the low farm incomes in this area and the possibility of better incomes "if fluid-milk markets were located in this area."[19]

Data from Cornell's Farm Management Surveys confirm this diversity of farming strategies and the link between diversity of practice and distinct economic networks, even at the county level. One study covered five areas "to represent the various systems of dairy farming in New York State."[20] Table 7.2 shows three of these areas. The data show that farms in specialized market milk Orange County had larger herds and higher capital investments. Orange County farms were also less seasonal and grew more high-protein crops such as alfalfa. Orange County farms sold milk primarily to Grade A markets in New York City.

St. Lawrence County farms represented the extensive milk "summer dairying" strategy in this survey. The farms surveyed in St. Lawrence County sold to cheese factories although four-fifths of the farms were also inspected to serve lower-priced Grade B milk markets. On average, St. Lawrence farms were smaller, with lower capital investments, much greater seasonality, and slightly lower yields per cow. The study showed that St. Lawrence County farmers surveyed fed much less high-protein feed and forage and depended more on pasture-feeding strategies. As a result, these farms had greater seasonality and somewhat lower production per cow, since pasture forage is seasonal and not as high in protein. The study admits "the importance of pasture in producing milk, both as a cheap source of feed for cows, and as a way to reduce the amount of labor spent in caring for the herd."[21] However, the author recommends against this summer-based strategy, because it does not use labor efficiently throughout the year. Therefore, the Cornell management advice to reduce seasonality and increase per-cow production, for the sake of labor efficiency, generally meant changing to a more intensive dairy strategy that involved less pasture and more feed crops.

Farms in Cayuga County were particularly small, but production per cow was high. Farms in this area were those "general/mixed" farms on better soil resources. These farms had the ability to grow the high-protein feeds necessary for higher milk production, but kept their herds small because they were also involved in the production of nondairy crops and poultry farming. These farms were practicing a more diversified dairy strategy that was possible with smaller herds because it included the cultivation of other crops.

American cheese production also tended to be located in low-resource areas, a fact made clear in the 1940 New York township-level data in table 7.3.

Table 7.2
Comparison of Dairy Farming in Orange, Cayuga, and St. Lawrence Counties, 1939–1940

County	Average Herd Size	Average Milk Yield per Cow (pounds)	Average Capital Investment	Seasonality[a]	Average Income[b]
Orange (winter dairy/ Grade A market milk)	28	6,635	$12,885	55%	$700
Cayuga (general farm/ Grade A and B market milk)	14	6,772	$10,466	50%	$468
St. Lawrence (cheese/ Grade B market milk)	19	6,517	$9,429	37%	$379

Source: Bierly, *Factors That Affect the Cost and Returns in Producing Milk.*
[a]Seasonality represented by the percentage of milk sold in winter months, October–March.
[b]Income calculated as farm receipts, minus farm expenses, minus interest—at 5 percent—on capital investment (to account for fixed costs).

Table 7.3
Percentage of Milk Plant Types Located in Poor, Fair, or Good Soil Quality Areas, 1940

	Soil Quality[b]		
Plant Type[a]	*Poor*	*Fair*	*Good*
Milk	12%	37%	51%
Grade A	7	40	53
Grade B	14	35	51
Cheese[c]	22	60	18
Cream	20	49	31
Butter	17	25	58
All Plants	18	45	37

Source: DuPuis, "The Land of Milk."
[a]Includes receiving stations, does not include urban processing plants.
[b]Soil quality areas defined according to author's determination of average soil quality in each New York rural township.
[c]Most New York plants produced American cheese.

As these figures show, industrial milk production developed as farms with the resources to intensify production forged linkages with fluid milk markets. More than half of all market milk townships in 1940 were located in good soil areas. Cheese factory farms were less dependent on high-quality resources. Only 18 percent of farms in cheese factory townships had good soils. The pasture-based dairy farming strategy was an adaptation to the quality of available resources.

In sum, cheese dairy farmers pursued a style of resource use that was made possible by the existence of a particular economic network and their links to that network. Cheese factory farmers sold their milk seasonally, produced it with little capital investment in technology, feed, or extensive housing, and sold it at a relatively low price to the local cheese factory. Because their feeding system did not require intensive growing of high-protein forage, which requires better soil resources, these farmers could make a modest living from comparatively poor soil resources. The location of cheese factories in 1940, relative to soil quality, indicates the tendency of cheese production to be located in these lower-resource areas.

MARKET MILK AND CHEESE FACTORY NETWORKS: AN INSTITUTIONAL HISTORY

To understand the linkages between farm strategies and economic networks, we must first understand the differences between market milk and cheese factory economies as institutional networks. Both the market milk and cheese factory economies developed as spatially and institutionally distinct systems over a long period of time. A history of the development of these two economies gives an indication of the distinct set of linkages involved in these two forms of production.

The cheese factory network is often presented as the earlier system of production from which market milk dairying emerged. In fact, fluid milk and cheese factory networks began and developed in parallel with each other. Before the development of an economy based on market milk and factory cheese, farm women made dairy products at home. Yet this was not simply subsistence production. "Farmhouse" cheese had been a globally traded product since at least the 1700s.[22] By 1850, American farmhouse cheese dominated world markets, and upstate New York was the center of this global dairy production system.[23]

In butter production, the farmhouse system survived into the twentieth century. Cheese production, on the other hand, rapidly moved to factories in the mid-nineteenth century. In 1851 Jesse Williams opened the first cheese factory in Rome, New York, less than a decade after Thaddeus Selleck started the first milk train from Orange County. The cheese factory and market milk production systems, therefore, coexisted for more than a century.

The parallel rise of market milk production for cities and factory cheese production for world export, all within a few years and a few hundred miles of each other, should be considered a story about diversity in the history of agrarian transformations. In particular, it says something about the slow transformation of city markets from swill to country milk. The reformers had to exhort farmers to provide milk to cities because farmers for many decades had viable alternative markets for their milk in the form of butter and cheese. The development of the first associated cheese factory the year before Mullaly's call for more milk for cities is an interesting example of heterogeneity in the organization of production and of markets.

Like their pasture-farming clients, cheese factories operated seasonally. Alfred Chandler, who writes extensively about the development of industrial production systems, has noted that large-scale production is dependent on a system that runs all through the year. Seasonal industries, in contrast, could not be organized along the lines of "modern" enterprises.[24] Seasonality of production on the farm and the scale of production in the factory were closely linked.

Cheese plants were therefore small. Yet there was another reason for the small size of these plants: transportation. Up until the 1930s, farmers tended to haul their own milk to the local plant, by horse and wagon over unimproved roads. There was a limit on the distance a farmer could reasonably haul milk, generally no more than five miles. The small cheese factories were therefore located at numerous country crossroads. Many small plants were dispersed over the countryside. Yet, unlike milk stations, cheese plants did not need to locate near large-scale transportation infrastructure, either railroads or, later, highways. They also were not limited in location to those places where railroads and highways could be easily built, namely, the valley areas.

Cheese factories, therefore, were a crucial institutional link in a less intensive, smaller-scale food provision circuit that had little to do with the institutional linkages necessary to bring fresh milk to the city. Local cheese factories were established in a wide array of institutional networks:

> As the factory system developed, a building was erected wherever and whenever there were a few farmers owning 200 or more cows within a radius of two or three miles. These factories were started in one of three ways. One way was by the farmers in association, either cooperatively or otherwise, furnishing the capital required. Another way was by cheesemakers who had served an apprenticeship in cheese making and were seeking an opportunity to organize a factory. . . . The third way was by promoters who started factories where farmers were slow in organizing. These promoters built factories, operated them for a season or two until the plant had secured a good patronage and then sold out.[25]

By the 1880s, merchants had purchased most factories from associations and they became privately owned.[26] Often, a merchant owned several factories in an area, as shown by the ownership pattern of factories in Cattaraugus County in 1906 (fig 7.1). Factory production remained decentralized because of the procurement problem: moving perishable milk long distances by wagon over farm roads was difficult, and production from old milk tended to affect the keeping quality of the cheese. Cheese factories therefore tended to receive milk from a procurement area of only a few miles' distance. Throughout the last half of the nineteenth century, merchants built more factories, generally where two roads met, engendering the name "crossroads cheese factories" for this type of plant.

The fact that railroads tended to be built in valleys also meant that fluid milk plants tended to be located in valley areas with better soil for intensive feed production. The railroads provided a convenient way of spatially demarcating the boundaries between the more and less intensive types of dairying in New York: "since the roads were poor and transportation by horse and wagon slow, it was only the farmers located near a railroad station who found the new outlet for fluid milk attractive."[27]

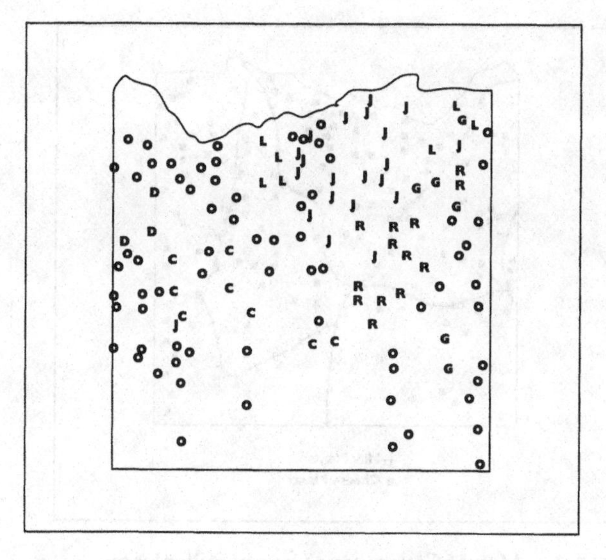

FIG. 7.1. Map showing ownership of cheese plants, Cattaraugus County, 1906. (Each letter represents one owner.) From DuPuis, "The Land of Milk."

In 1861 a New York farm magazine, the *Rural New Yorker*, described the separation of these two systems in geographical terms: "Five miles by wagon, 70 miles by rail is as far as milk can be carried. Thus a railroad can draw milk from a strip of territory 10 miles wide and 70 miles long."[28]

A map of Cattaraugus County milk plants and railroads in 1906 (fig. 7.2) shows this spatial separation between fluid milk and cheese production regions. It shows that milk plants were located only near railroad lines while cheese plants were dispersed through a wide area. Wagon transport of milk meant that fluid milk production was restricted to an area no more than a few miles from the railroad. A map of Allegany County plants and railroads in 1938 shows fewer cheese plants but the same spatial relationship (fig. 7.3). These maps show that for many decades, market milk plants followed rail lines, while cheese plants were widely dispersed through the countryside. As refrigerator cars came into use in the 1880s, farmers along the length of the railroad's line, but not outside the ten-mile corridors, could potentially

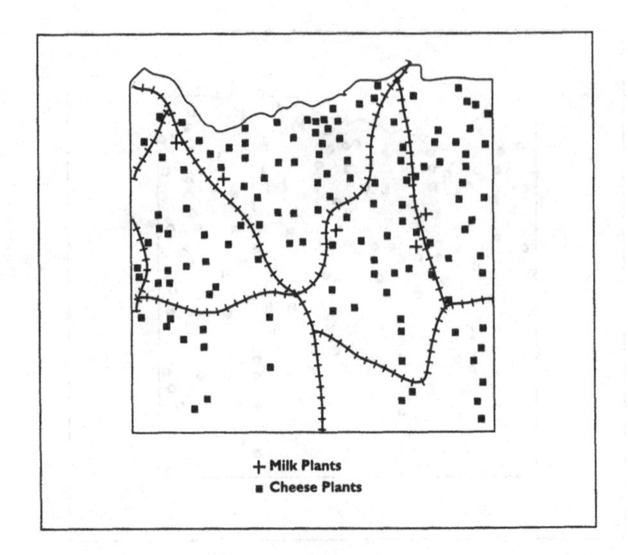

FIG. 7.2. Map of Cattaraugus County milk plants, cheese plants, and railroads, 1906. From DuPuis, "The Land of Milk."

serve city milk markets. With the increase in the demand for milk, farms in the far northern regions of the state became part of the fluid milk market, as long as they were within the ten-mile milk train corridor. As early as 1879, farms over 250 miles away served the New York City milk market. By the turn of the century, milk trains soon reached over four hundred miles away from the city for milk supplies. In addition, until 1897, hauling rates were uniform over the entire length of the line, thereby encouraging farmers at the extreme ends to ship milk to city markets.[29] Yet the width of this city milk strip expanded little.

The cheese factory network, therefore, not only differed from the market milk network in the nature of dairy practices, resource use, and infrastructure linkages, they were also, until the 1930s, spatially distinct. The physical limitations of transportation technology created relatively stable spatial barriers between the fluid and manufacturing milk regions for nearly a century. The limitations on farmers' ability to get milk to the train station helped to create a spatial boundary between fluid and cheese milk producers. These limitations contained much of the "leakiness" between these economies, that is, it protected both economies by keeping the two systems of milk production out of competition with each other. This spatial barrier, in other words, also acted as an economic barrier, which protected the two economies as distinct forms of production. As the next chapter will show, changes in transportation put these two systems in competition and threatened the viability of both networks.

In many ways, this system was perfect for New York state's geographical landscape. As noted earlier, much of the state, especially the "southern tier," is hill land crossed with many narrow valleys. These valleys had the better farmland, which could grow the winter feed necessary for yearlong milk production. These level valleys were also suitable for the construction of train tracks.

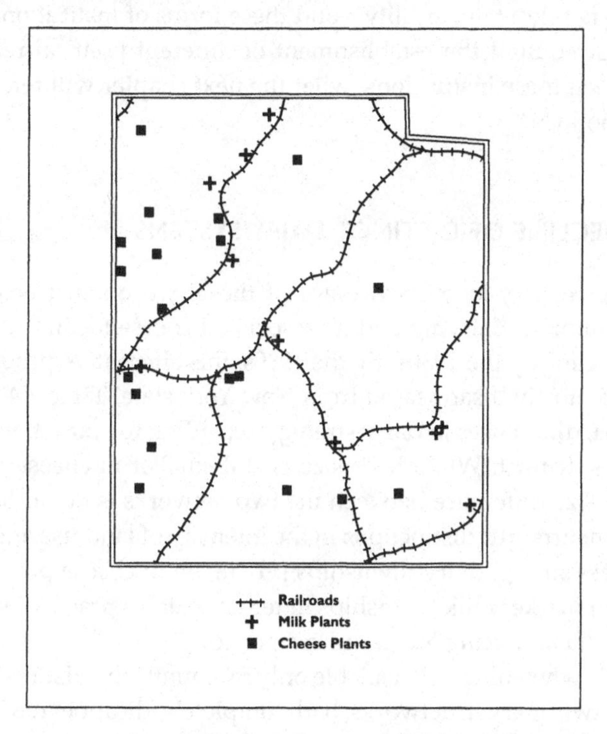

FIG. 7.3. Map of Allegany County milk plants, cheese plants, and railroads, 1938. From DuPuis, "The Land of Milk."

The proportion of milk that went to fluid sales steadily grew. In the country as a whole, sales of milk for fluid use accounted for only 6 percent of total milk production in 1849 but had risen to 28 percent by 1929.[30] In New York, the change was even more dramatic, rising from only 9 per cent of the total production sold as fluid milk in 1864 to 72 per cent in 1927.[31]

The growth of the market milk system involved bringing larger and larger amounts of this fresh, perishable product longer distances into urban areas. This required a substantial amount of control over the production and distribution of the product. Cheese milk farmers and processors had no need for such a system.[32] The growth of cheese production simply involved establishing more factories. Therefore, each of these networks did not simply represent a "market" link between farmer and processor. These networks were also embedded in other institutions—family, state, locality—and these forms of institutional embeddedness required the establishment of different political relationships between these institutions, what the next chapter will refer to as "forms of politics."

THE DECLINE OF DISTINCT DAIRY SYSTEMS

Despite the century-long coexistence of these two distinct economic networks, summer dairying and its associated cheese factory network began to decline by the 1930s. By the 1980s, this alternative production system had mostly disappeared from New York state. Table 7.4 shows that by 1964, differences in dairy strategy according to market type had become less distinct. While herd size is still smaller in cheese factory townships, size difference between the two networks is not as large as in 1940. Concurrently, distinctions in the intensity of land use and feeding were less strong; dairy townships proximate to cheese plants converged with market milk townships in terms of dairy practices such as feeding alfalfa and using less farmland per cow.

By 1987, when data are available only by county, the distinction between the two market networks had completely disappeared. Using county-level data, while allowing for less distinct variation between smaller geographical units, also reflects the greater movement of milk and consequent overlap of milk between markets, which in fact makes a spatial distinction between markets no longer analytically possible by

Table 7.4
Dairy Strategy by Market Network, New York Dairy Townships, 1964, and Counties, 1987

Strategies	Market Network	
	Fluid Milk	*Cheese Factory*
Average Herd Size (cows per farm)		
1964	32	28
1987	59	55
Average Intensity of Land Use (farm acres per cow)		
1964	9.8	10.9
1987	12	13
Average Feed Intensity (alfalfa acres per cow)		
1964	0.90	0.76
1987	1.3	1.2

Source: DuPuis, "The Land of Milk."
Note: Dairy townships are defined as townships with more than five hundred cows.

this time. Today, linking data on farm strategy with local plants at the township or even county level is no longer possible, since cheese production as a distinct and local sector no longer exists in New York state. In part this is due to the greater movement of milk; dairy plants are no longer local. However, the disappearance of the summer dairy/cheese factory network also has to do with the transformations that have taken place in the dairy manufacturing sector, which will be a topic of chapter 8.

These data record the disappearance of summer dairying as a distinct dairy strategy in New York state. Distinct differences in dairy practices now aggregate at a much broader regional level (West, Midwest, Northeast).[33] Overall, while dairy farming has not achieved the industrial vision of perfect farming, production practices across farms have become significantly more intensive and less diverse. Today, fewer dairy farms with more cows produce more milk than ever before.[34] Between 1950 and 1990, the amount of milk produced per cow rose from less than 5,500 pounds per year to 14,500 pounds per year, due to the intensification of dairy practices.[35] Dairy farms are getting larger and the number of farms reporting dairy cows has dropped from over 4.5 million in 1900 to 155,000 in 1992.[36]

Unlike farms in the Northeast and Midwest, western dairy farms closely fit the neoclassical ideal of industrial efficiency. "Dry-lot" farms in the West buy most of their feed and have the latest milking technologies. Western dry-lot dairy farmers are farm managers and labor is carried out mostly by employees.[37] These farms are outcompeting other dairy regions and accounting for a larger share of U.S. milk production, rising from 15 percent to 24 percent between 1980 and 1994.[38] From the neoclassical point of view, agricultural production in the past may have been diverse, by region and from farm to farm, but the economic structure of agricultural production is increasingly "converging" to become the more perfect, efficient farm of the future.[39]

Areas adapted only to pasture-based dairying in New York have moved out of farming altogether, leaving the picturesque decaying barns and overgrown fields visible in hill areas throughout the state, a landscape that has become the nostalgic subject of many regional artists.

CONCLUSION

Despite its fresh, untouched appearance, fluid milk is an industrial food. This is at first thought surprising, since one always associates industry with processing. However, the amount of modern technology and modern business organization involved in taking a food utterly prone to contamination and delivering it safely and daily to consumers, often hundreds of miles away, is more intensely industrial than the dairy manufacturing processes that require no specialized transportation organization.

According to the industrial vision, only one dairy system was perfect: the industrial winter dairy system. From the perfect story perspective, the decline of the cheese factory/summer dairy system is unproblematic and automatic, in tune with the general progress of the dairy industry toward greater productivity and efficiency. As with consumption, this narrative of inevitability limits the questions we can ask about milk production.

In fact, as the next chapter will show, the decline of this economic network was neither automatic nor unproblematic. The logic of industrialization—which the perfect story mirrors—was certainly at work, but this seemingly inevitable force received significant help from agricultural experts. Whether summer dairying or mixed farming survived depended on human decisions that emphasized the superiority of large-scale systems. As chapter 8 will show, these reasons were often political.

While dairy experts have been unable to remake farms entirely according to the industrial vision, increasing production per cow remains the primary focus of most dairy experts. The economists' industrial vision of neoclassical factor efficiency is so powerful because it relates to the world in three different ways. First, it is a *mirror,* reflecting the real forces of modernization, globalization, and the convergence of production processes to the most efficient form. Second, however, the economic story of efficiency is also a *mask* that covers up the true diversity of production practice that has existed and still exists today. Finally, this story, when used to create agricultural policy, becomes a *hammer,* a self-fulfilling prophecy.

As a *mirror,* the perfect story of economic efficiency simply reflects the actual happenings in the world today, the process generally known as "modernization." We do live in a world in which the "logic" of modern life pushes us toward market-based economic efficiency. This is the story of the "invisible hand" that Adam Smith celebrated, the "iron cage" sadly predicted by Max Weber, and the "monopoly capitalism" that Karl Marx hoped would eventually destroy itself. Dairy is no exception. Taking Adam Smith's view, neoclassical economics paints a picture of progress that celebrates dairy herds becoming larger, cows becoming more productive, and regions where dairying is capital- and labor-intensive winning out over regions where dairy farming is primarily family-based. While sociologists have been more ambivalent about this force, they have recognized its power to transform society.

Yet the perfect story is, at the same time, a *mask*. In Walter Benjamin's words, historians who describe the history of progress make time "empty" and "homogeneous."[40] The neoclassical storyteller ignores or trivializes the diversity of practice that the progress story cannot explain, or this diversity is trivialized as an earlier and inferior practice. The perfect story masks the diversity of practice—the other ways of doing things—in the past and minimizes the diversity of practice that remains today.

The perfect story of farming tends to ignore this diversity of practice and the existence of multiple "styles," "strategies," or "systems" of farming. The perfect farming story, as with the perfect consumption story, relies on a concept of time unilinear and a concept of space as homogeneous. Change is impersonal and inevitable and the ultimate goal is a single, uniform, universal way to produce food. As with the story of perfect consumption, challenging the destiny story of industrial agriculture requires challenging the notion of perfect production. To do this requires questioning the inevitability of large-scale forces and impersonal motors of change. Instead, it is necessary to emphasize the social relationships, the linkages and the human decisions that made our food system what it is today.

Contemporary scholarship challenges the notions of the universal, the essential, and the perfect, to readmit explanations of economic change that give a strong role to social and political context. From this perspective, we can look beyond the idea of family farmers as the exception to the rule of economic industrialization and look at particular dairy practices as linked to particular historical, institutional, and organizational communities that lie beyond the farm gate.[41] How these alternative communities of practice survive becomes an interesting question, not simply because we might favor diversity, but because we want to understand how society really works behind the mask of the perfection story. The less perfect story of dairy production recognizes the diversity of agricultural practice as a part—although not the only part—of the story. Yet it is also important not to make diversity into the subject of a new story of perfection. The less perfect story is not an ethic in itself; it is a way to look at and explain social relationships in ways that take diversity into account. A story that emphasizes diversity of practice has room for a diversity of ethics and politics.[42]

Current research in Europe shows that dairy farmers there continue to pursue a variety of farming strategies made possible through their links with unique regional production economies. For example, the production of regional cheeses, such as Parmesan, involves institutional linkages with organizationally distinct systems of food production and consumption.[43]

Researchers from the Wageningen Agricultural University in the Netherlands are central to forging this new approach to the question of agricultural diversity. Stating that "[d]iversity is one of the main features of European agriculture," Wageningen sociologists argue that "[a]ny European perspective on rural development must be grounded on the recognition of such diversity and must necessarily build upon it in order to maintain the agriculture required by Europe's peoples."[44]

From this perspective, agricultural diversity is not a remnant of past traditions but a resource in itself, a form of "endogenous development"—what from the feminist perspective on breast-feeding explored in chapter 3 might be called a "community of practice."

Whether described as endogenous networks or communities of practice, these forms of local social capital are resources internal to a region, as opposed to resources gained from outside.[45] In local endogenous networks, particular farming styles "represent specific connections between economic, social, political, ecological and technological 'dimensions,'" and each style contains "a specific co-ordination of the domains of production and reproduction, the domain of economic and institutional relationships and the domain of social (i.e. non-commoditized) relations."[46]

Finally, and most important, the industrial story of perfect farming is a *hammer*. By proclaiming the triumph of perfect, efficient, industrialized farming and ignoring the actual diversity of farming systems and farm practices, the perfect story becomes a tool that creates its own future. When a pro-industrial economist tells the story of perfect production, people listen closely, even if they don't necessarily like the punch line. Based on this story, policy decisions often emphasize ways to make the transition out of farming easier for smaller farmers, or ways to move more "resources" out of agriculture.[47]

The creation of milk as a perfect, essential food created the impetus for a large-scale urban intervention in the dairy production system. The cheese factory summer dairy network was politically independent of this industrial bargain. This system of dairy farming, like most previous dairy systems, revolved around preserved dairy products. The intense city-country linkages that characterized fresh milk markets were simply unnecessary.

The new emphasis on agricultural diversity has led to a rethinking of the relationship between farmers, researchers, and consumers. Sociologists of agriculture working for the European Union have realized that this new approach to agricultural development must differ substantially from the old policies based on the industrial vision. In particular, they are arguing for a move away from the "enlightenment" model of industrial agriculture, which depends primarily on the input of technology and knowledge from outside. Instead, policies encouraging diversity of practice seek to give the farmer "an essential role in the *valorization* of endogenous resources of the area where he exercises his activities." This new approach to rural development emphasizes "the identification of endogenous potential" in these areas "rather than analyzing the short-comings and constraints of these regions."[48] Instead of developing agriculture in a region through the application of a standardized "package" of high-tech inputs, the new approach "relies on the cultural identity and the specific patrimony of the areas concerned (landscape, craftsmanship, history etc.). This means that solidarity in favour of rural areas must be organized in deference to their diversity."[49] From this perspective, cultural identity and diversity are key resources in a particular regional economy.

What also becomes clear is that market milk farmers were not only in a different economic position from cheese milk farmers, they were in a different *political* position as well. As we will see in the next chapter, for market milk farmers, farm radicalism—milk strikes and market-wide collective action—seemed to hold promise. Yet, as an examination of Wisconsin and California dairy policies in chapter 9 will show, manufacturing milk farmers needed different policies to remain viable, ways not available to the farmers of New York state.

8

Crisis

The "Border-Line" Problem

ON A JULY MORNING in 1939, a crowd of dairy farmers gathered near the gate of the Sheffield Farms plant in Heuvelton, New York, at that time the largest milk receiving plant in the world.[1] As members of the Dairy Farmers Union, these farmers had voted the night before not to deliver their milk to the plant. They formed a picket line at the plant gate as they tried to convince other farmers arriving with their loads of morning milk not to deliver as well. The strike closed the Heuvelton plant for over a hundred days.

The Heuvelton plant was four hundred miles from New York City, but most of its milk went to the city's fresh milk market. Since the 1920s, companies with "classified" pricing plans paid for milk according to its "class" of use; fluid market milk received the highest price. The same classified pricing system had become part of the recently established New York City milk market order.[2] The order made classified pricing mandatory for those farmers who served the city fluid market. The right to set prices had already survived a series of court challenges. Now, supported by federal market order legislation, market orders became the established system of fluid milk pricing for the rest of the century.

Dairy farmers were not entirely satisfied with the prices set under the order. In particular, farmers who were not members of the two major dairy cooperatives—the Dairymen's League or Sheffield Farms—believed that the prices set favored cooperative members.[3] However, dairy farmers in Heuvelton were not striking about the order. They had another problem: Sheffield was about to close this plant and twelve other plants in the northern New York counties, leaving many farmers in this area without access to the New York City milk market. Losing access to this market meant having to accept a much lower "manufacturing" classified price for the milk these farmers produced.

By 1939, the milk strike was an old and well-used tool for New York state's dairy farmers. In fact, the first farmers to strike for higher milk prices were farmers in the Orange County dairy region Robert Hartley had visited a hundred years before in search of country milk. By the 1880s, city dealers had formed a price-fixing cartel called the New York Milk Exchange, and as a result, milk prices "began to fall steadily."[4] In response, the farmers formed the Erie Producers Milk Association in 1883 and withheld their milk from the New York market. In an action prescient of decades of conflict to come, the dealers "began immediately to take steps to prevent a recurrence of this situation by establishing other sources for their milk," employing new systems of refrigeration that kept milk longer.[5]

> The dealers' response made the farmers' cooperative associations comparatively ineffective, since the new dairymen coming into the field welcomed the [dealers] as the Orange County men had done earlier. In order to control the milk trade, the farmers had to remain united and to do this they had to get the new men into their ranks to prevent the destruction of their organization. This they could not do and thus soon fell to quarreling among themselves, distrusting each other, and criticizing their leaders.[6]

The happenings of 1883 echo down to the global agricultural political economy of the present day. Global "multiple sourcing" as a business strategy involves the structuring of production so that a multinational firm can buy a necessary product—whether orange juice or car components—in more than one place.[7] The business literature praises such a strategy as giving the sourcing firm less "risk" and "uncertainty" and enabling regions to gain new sources of competitive advantage.[8] Yet the risk multinationals face is the rise of prices paid through the political organization of producers, and the competitive advantage nations receive depends, to a significant extent, on their ability to quell producers' price demands. Global multiple sourcing gives multinationals the ability to make their production regions compete with each other for the opportunity to supply the firm with this product. This competition often involves a "leap for the bottom" in terms of wages, child labor laws, environmental regulations, and other working conditions.[9]

In the interest of ameliorating these conditions, social activists have begun to agitate for greater worker power through the creation of global coalitions of workers.[10] This is a major political challenge, since multiple sourcing tends to create its own form of politics. Rather than inspiring collective social movements of workers making claims against their employers—a class form of politics—multiple sourcing inspires a "sectionalist" politics induced by regional competition. Multiple sourcing reduces regional development strategies to a "chasing smokestacks" race between regions. The interregional competition caused by multiple sourcing undermines producer solidarity, leading classes that potentially share political interests to "quarrel among themselves." In other words, multiple sourcing—on a regional or global level—creates a particular form of politics that tends to frustrate class-based collective resistance.[11]

Political sectionalism is not new. In many places, regional political solidarity has followed ethnic and religious lines, as the history of the Balkans has shown. While sectionalist politics is often based on this ethnic or religious competition, in some cases the competition has been economic. Political scientists have observed that post–Civil War sectionalist politics in the United States has tended to follow economic rather than ethnic or religious lines.[12]

The 1883 milk strike and the dealers' strategic response show that this form of politics is not limited to either the present time or to global spaces. While the politics of milk have been inherently spatial, the regions involved have not generally been worlds apart. For most of a century, milk's geographically restricted production territory enabled milk companies to play different production regions off each other simply by signing on farmers in the next county.

Studies of dairy politics have not taken this regional competition into account. Political scientists who have looked at the politics of dairy have focused on the congressional "dairy bloc," an interregional coalition that has for many decades forged a powerful political compromise with other regional commodity blocs such as corn and cotton. It is the interregional horse-trading of these three blocs that has kept the U.S. agricultural subsidy system in place.[13] From this national perspective, dairy producers function under a "corporatist" form of politics, in which producers and milk companies are each organized into national federations representing their conflicting perspectives, which have been mediated by the state.[14] Yet regional competition *within* the industry continues to the present day.

Histories of dairy collective action in New York agree about one thing: attempts at collective action by the state's dairy farmers ended in failure. Most blame the statewide market cooperative, the Dairymen's League, in some way. For some, the problem was monopoly control by top Dairymen's League officials.[15] For others, the problem was bad decisions made by these officials.[16] The common theme in both stories revolves around what Spencer and Blanford term "the problem of disunity."[17] Accordingly, most stories follow the description of rapid success in collective organization against milk dealers' monopolistic activities, followed by a decline into divisive politics that weakens the movement.[18] In fact, as a comparison with Wisconsin and California will show, dairy farmer disunity in New York state was the product of political conflicts in the industry that went beyond the decisions of the state's major dairy cooperative. The form of the cooperative itself was a product of New York's unique form of politics.

Another group of actors influenced the border politics of milksheds. Over and above the activities of larger companies, smaller milk dealers could undercut prices by signing on new farmers. In some cases, smaller dealers attempted to underprice larger dealers by selling lower-quality milk. As a result, the New York City Health Department instituted a licensing requirement for dealers in the 1890s. However, even without lowering quality, smaller dealers with licenses had a competitive advantage: they could buy just the amount they needed to sell, whereas larger companies had to deal with the problems of surplus milk during those times of year when milk production was high.[19]

The geopolitical gamesmanship between and among dairy producers and milk dealers tended to backfire on the entire industry. An overexpansion of the milkshed would lead to an even greater "surplus" of milk in the New York City market and increased price competition at the retail level. This concomitantly led larger firms to move producers at the outer edge of the milkshed out of the market in order to raise milk prices. But these producers did not go easily, as indicated by the northern New York farmers' reaction to the closing of the Sheffield plants. The loss of access to the higher-paying fluid milk market resulted in a decision "to strike the twelve Sheffield plants."[20]

The milk strikes of the 1930s therefore involved groups of farmers with political interests that were not identical. First, farmers on the edge of the milkshed, such as those at Heuvelton, were responding to their loss of a market. Sheffield was closing the Heuvelton plant in order to deal with the surplus milk problem that was lowering milk prices. Second, farmers throughout the milkshed reacted to the actual drop in prices, at times declining as much as 50 percent.[21] Needless to say, since restricting market access was one way to deal with the drop in prices, farmers dealing with low prices did not necessarily feel in complete political solidarity with those who had lost access to the market. In other words, as in the smokestack-chasing regions of the globally sourced economy today, regional competition rather than cooperation often seemed like the better strategy, especially in situations where market demand was stagnant. As a result, while populist accounts portray small independent farmers in struggle against the large dairy firms,[22] in fact, much of the struggle took place as regional sectionalist politics between farmers.

The 1939 Dairy Farmers Union strike was relatively nonviolent compared to previous strikes. For example, the 1933 strikes had gotten particularly out of control. These strikes "possibly . . . brought New York State closer to martial law than at any time since the Revolutionary War."[23] Five days of *New York Times* headlines during the 1936 milk strike also illustrate the extent of disruption:

August 5: "Strike Violence"
August 6: "Rioting—Violence Continued"
August 7: "Strike Violence"
August 7: "Governor Orders Appointment of Deputy Sheriffs"
August 8: "Milk Dumping—Violence"
August 9: "Strike Violence—Shooting"

While farmers also directed violence against the large milk companies, strike accounts show that much of the violence was directed at other dairy farmers attempting to continue deliveries of milk to dairy plants.[24] Dairy Farmers Union strike actions included a mock funeral with two coffins, one for the large milk companies and the other for the Dairymen's League.[25]

For the most part, cheese factory farmers did not use milk strikes as a way to raise the prices they received.[26] Cheese factory farmers couldn't leverage their production for this purpose, since cheese was not a fresh food that needed to be supplied to customers every day, nor was it considered an essential food. In contrast, three companies—U.S. Dairy, Borden's, and Sheffield Farms (National Dairy, later Sealtest/Kraft)—controlled two thirds of the fluid milk market.[27] As these companies grew to become modern industrial giants in many of the country's large cities, the relative power of seller versus buyer became increasingly unequal. As far back as 1883, milk dealers—organized in a dealers' "exchange"—had great ability to control the milk market through a multiple sourcing strategy. The rise of milk as a fresh, indispensable food created the "milk strike," the multiple sourcing strategy, and the resulting sectionalist form of politics.

POLITICAL CRISIS AS BOUNDARY CRISIS: THE ARRIVAL OF TRUCK TRANSPORT

As noted earlier, New York state's dairy landscape was riddled with geographic features—both natural and historical—that affected the way dairying was spatially organized. These features made some farms better located in relation to the market, but good location had to do with more than simply being close to the city. The farm also had to be located close to the railway, in particular, close to one of the country milk stations that collected and stored milk for railway transport. This locational factor was critical in determining whether or not a farm could participate in the milk market. For many decades, transportation limitations created a physical limit on the milkshed boundary, protecting the integrity of the two markets.

In the 1930s, the advent of truck milk hauling eroded the geographical boundaries that segmented dairy production. Truck transport rapidly expanded as a form of milk transport during this decade. Accounting for only 10 percent of transport in 1931, trucks moved more than half of New York City's milk by 1938.[28] As a result, the expansion of the New York milkshed after 1940 was not so much an extension of the outer boundaries of the shed but a more intensive procurement of milk within the shed that had formerly gone to other uses.[29] This more intensive coverage was to some extent made possible by the extension of milk trucks into farm areas previously without access to train lines.

With the arrival of milk transportation by truck, the physical separation between cheese-producing and fluid milk–producing regional networks began to break down. Farmers outside the railway corridors now had access to the fluid milk market, leading to increasing instability in the boundary line between the two dairy systems. This brought hill cheese and butter farmers—with their lower costs and seasonal system—into greater competition with market milk farmers—with their need for higher prices to make a return on their higher investments, and a need for stable prices over the year to make a return on their winter production emphasis. Instability in prices led to uncertainty for farmers as to whether or not they would make an adequate return on their investments, particularly for farmers making higher investments to serve the fluid milk market. While it is easy to point to the Depression as the cause of the dairy crisis in the 1930s, the drop in demand due to less consumer buying power was simply an exacerbation of this more fundamental change in the industry.[30]

The extent of instability in a milkshed therefore depended on the "leakiness" of the boundary between the two dairy networks. If the fluid milk producers demanded higher prices, or if milk demand or fluid milk production varied temporarily, milk dealers would try to expand their procurement system into the manufacturing areas, the cheese and butter production networks. During the "flush" season, milk from manufacturing areas was most likely to come into the fluid market, exerting a downward pressure on fluid milk prices.

Some economists acknowledged this problem. As milk prices dropped with lower demand during the Depression, hill dairies, "in desperate need of income to meet fixed expenses such as taxes and interest," searched for an access to the fluid milk market.[31] It was obviously tempting for dealers to buy this cheaper supply and sell it at a lower price than the competition. The agricultural economist M. C. Bond, writing at the time, described the political instability of "new and increasing 'border-line' problems": "Since transportation has had so much to do with the area of a milk shed, it is not surprising that important changes have occurred in recent years as more and more milk has been moved by truck."[32]

The advent of truck hauling of milk led to pressures to spatially restructure the entire New York milkshed. The spatial organization of the milkshed in New York and many other peri-urban milksheds across the country began to change significantly during this period. Rather than being organized in five-mile swathes on either side of railway tracks, the market began to be organized more as a ring around each city market.

While the ring form of organization made sense in a truck-transportation–based dairy market, moving from the railroad strip to a ring-shaped milkshed involved serious political problems for those who now found themselves at the "edge" between markets. Market edges in the concentric ring system were not set by physical constraints as they were during the railroad strip period. The question under this form of spatial organization became, How far was too far away to serve the fluid milk market? Producers who were far beyond the fluid milkshed understood the prices they faced in the manufacturing market. They could make decisions about their farming strategies within a stable market environment. Producers close to the city were solidly within the borders of the fluid milkshed and therefore certain of their farming strategies. Their economic return, however, was less certain, depending on the size of the milkshed as a whole, in particular the extent to which it included surplus milk, which affected the "blend price" the farmer received for milk.

Differences in natural resources exacerbated this picture. As noted above, New York state was well suited to the train strip form of spatial organization, since the train tracks traversed the valleys where intensive dairying could take place. As the truck transportation system developed, more farmers on lower-quality land could potentially enter the fluid milk market. Yet much of this land was not well suited to grow the crops—corn and alfalfa—necessary for winter feeding. These farmers were therefore less likely to reduce their seasonality, and thus the problem of surplus milk in the summertime was exacerbated.

Truck milk transportation was also more highly dependent on public investment compared to milk train infrastructure. Large-scale plants needed both access to large amounts of milk and an efficient system for moving that milk. Larger plants also needed a well-developed road infrastructure to handle trucks all the way to the farm gate. Therefore, the development of a large-scale dairy system was dependent on public investment in improved roads in dairy areas.

PERFECTING SPACE

The result of these political pressures was several decades of economic contention, in which the milk strike became the ultimate tool. Economists refer to this situation as a "disorderly market," and the market "order" legislation, part of the Agricultural Marketing Agreements Act of 1937, was established to do just that: create orderly markets. Therefore, for government economists, the establishment of the perfect food and the perfect farm also led to the envisioning of a perfect, orderly market.

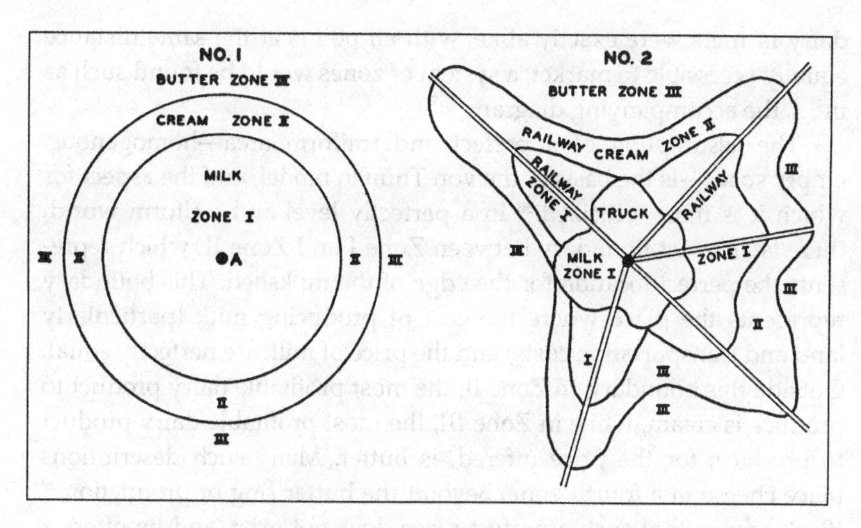

FIG. 8.1. (*left*) John Black's diagram of the ideal milkshed. From Black, *The Dairy Industry and the AAA*. FIG. 8.2. (*right*) John Black's modified diagram of a milkshed, accounting for transportation infrastructure. From Black, *The Dairy Industry and the AAA*.

The Harvard University economists John D. Black and John Cassels were the first to lay out a vision of orderly marketing for the dairy economy, a vision that depended largely on the economic geography of Johann Heinrich von Thünen.[33] Von Thünen's masterwork, *The Isolated State*, describes the spatial and economic relationship between a city and its surrounding countryside. To von Thünen, the spatial layout of agricultural production around a town center was a product of the transportation cost of the various commodities provided to the center. As a result, von Thünen argued, the first "ring" of land around the city produced crops that were perishable, expensive to ship, yet had a high value and were therefore capable of producing a profitable return on high-priced land. One of these crops, according to von Thünen, was milk, a product that was also high-value and which, due to its perishability, needed to be produced close to its market.[34]

Black is clearly referring to von Thünen when he lays out the assumptions for his description of the "ideal" milk production area around the city, as he illustrated in fig. 8.1: "If the city were seated in the midst of a perfectly level and uniform plain, and all dairy farms and dairy farmers were exactly alike, with all points at the same distance equally accessible to market, a system of zones would be found such as in . . . the accompanying diagram."[35]

The assumption of a perfect and uniform area—homogenous, empty space—is the basis of the von Thünen model, and the aspect for which it is most criticized.[36] In a perfectly level and uniform world, there is an exact boundary between Zone I and Zone II, which represents the perfect location for the edge of the milkshed. This boundary represents the place where the cost of producing milk (particularly land and transportation costs) and the price of milk are perfectly equal. Outside this boundary, in Zone II, the most profitable dairy product to produce is cream, while in Zone III, the most profitable dairy product to produce, for the price offered, is butter. Many such descriptions place cheese in a fourth zone, beyond the butter ring of production.[37] Black admits that such a perfect place does not exist, and he offers a "modified" scheme, showing the effect of train and truck lines and topography on the boundaries of the various zones (fig. 8.2). Black notes that as consumption in the city grows, the milkshed extends outward, since there is demand for more milk. An increase in demand would result in a higher price, leading more farmers at the edge of the outside ring to change from cream to fluid milk production.

But, Black posits, what if the price of fluid milk was raised arbitrarily, say, by government fiat? This would also prompt more farmers to enter the fluid milk market, but the result would be a surplus of fluid milk and a lower price for the product. Some of this milk would have to be sold outside the milkshed, back into the cream or butter markets, at the cream or butter price. Assuming that all farmers in the milkshed were pooled and therefore taking the same price, the price to the new farmers coming in from the cream zone would rise, but the price to the farmers who had been producing for fluid markets all along would drop. Another possible strategy would be to increase prices but find some way to exclude Zone II farmers from the fluid milk market. As the cross-state comparison in the next chapter will show, the outcomes of milkshed boundary struggles have involved the incorporation of a more inclusive, lower blend-price strategy that includes more farmers or a more exclusive strategy that involves restricting market access but gives a smaller number of farmers a higher price.

The institution of market orders for milk meant that the milkshed boundary, which was no longer subject to the physical restrictions of transportation, was to be tamed by the laws of economics. Yet in fact, the setting of the boundary remained open to political pressures. In times of decreasing demand, the market boundary became particularly contentious. During these periods, pressure increased to close plants at the outside edge of the fluid milk zone, to slow the precipitous drop in prices to all farmers serving the fluid milkshed. The boundary struggle involved a question of who would take the "pain" of a drop in market demand, farmers on the outer edge of the milkshed boundary, or all farmers in the milkshed? It is exactly this situation, the drop in milk demand during the Depression, that led to the moving of the twelve Sheffield Farms plants from fluid to manufacturing. Yet the organized farmers of northern New York did not accept the notions of market perfection through geography, or the idea that they were the ones who needed to suffer financially for the milkshed as a whole. In the Depression decade, with dropping demand, changes in transportation, and increasingly desperate farmers, the struggle over the extent of the fluid milkshed became known as the "border-line problem."[38]

Economists tend not to talk about political struggles. Instead they talk about market "instability." Accordingly, dairy economists' discussion of the boundary problem focused on the creation of "orderly markets." An orderly market, in fact, is one in which supply and demand are successfully met without political disruption. It is based on the neoclassical notion of perfection as the elimination of politics. The main issue from this perspective is not farmer unity, but the discovery of the right price. Yet the neoclassical notion of orderly markets can tell us something about the unity problem as well. Dairy economists' picture of "instability" is in many ways more complex and more interesting than simply the question of unity and disunity. From this perspective, the USDA dairy economist Alden Manchester has described how fluid milk markets became "inherently unstable":

> In the beginning, flat pricing was universal. Dealers paid producers one price for all milk, the price being sufficiently above that paid by manufacturing plants to induce producers to incur the expense of meeting sanitary standards for the market in question. The problem was to limit milk production and receipts to the quantities needed by the dealers, especially in the flush season. Dealers could add producers in the short season and cut them off in the flush.[39]

Sociological analyses of farmer disunity usually talk in populist terms about farmers against agribusiness or against larger farmers and their major representative association, the Farm Bureau.[40] Dairy farmers in milksheds where milk companies were practicing multiple-sourcing strategies, however, were divided as much by sectional differences as they were by differences in social class. Producers that tended to be added to the milkshed and "cut off in the flush" tended to be those at the outer edge of what dairy economists have called the "middle ring" of a milkshed. Those in the middle ring between established market milk farmers and "outer-ring" manufacturing milk farmers bore the highest risk in relation to market prices. A dairy farmer's political relationship to other dairy farmers therefore depended to a great extent on the geographic position of his or her farm in relation to the entire milkshed.

RING POLITICS: THE SECTIONALISM OF MILKSHEDS

The spatial incongruity of interests led to a form of "ring politics" in which farmers in each ring of the milkshed pursued a separate set of political interests. Discussions of the "disunity" problem have centered mainly around the inability of the Dairymen's League to attain and maintain the membership of farmers in the milkshed. While the league's 100,000 members represented more than three-quarters of all New York City milkshed farmers in 1922, membership had declined to only 33,000 by 1927.[41] Critiques of league decisions focus on the extent to which league officials signed up farmer members beyond what was necessary for the supply of fluid milk to the market. Before market orders established the market-wide sharing of a single price for milk, the price a farmer received had to do with who he or she sold to. Since price was usually a "blend" of the higher price of milk sold to fluid markets and the price of "surplus" milk that had to be used for manufacturing, farmers who sold primarily to the fluid milk market got a higher price. On the other hand, a market-wide cooperative with too many farmers would have to sell more of its milk at the manufacturing price.

Inner-ring farmers, with their easy access to high-paying city milk markets, were less likely to belong to the statewide cooperative, the Dairymen's League, than middle-ring farmers.[42] Farmers established this cooperative in 1907, but it did not take on a significant, statewide role until John Dillon, the commissioner of Foods and Markets, Dillon, helped establish the league as the state's major cooperative during the milk strike of 1916. During this strike, the most successful in the history of New York state dairying, the Dairymen's League took the lead as the major representative of farmers' interests.[43]

League officials tried to solve the blend-price problem by purchasing and running league manufacturing plants. This enabled the league to find a place for surplus milk that could result in a higher price for members. But investment in manufacturing facilities—generally not crossroads cheese factories but large-scale facilities—was expensive, and the League took the cost of these investments out of members' paychecks. As a result, the price that each member received—a price that took into account the cost of manufacturing facilities plus the lower price received for the milk that went into manufactured products—was lower than the prices received by other market milk farmers. This meant that farmers with other access to fluid milk markets eventually abandoned their membership in the league. League members increasingly were farmers with no other access to the milk market, seeking somewhat higher milk prices than the manufacturing alternative.[44]

In an attempt to deal with the surplus milk problem, fluid milk processors, particularly the Dairymen's League, increasingly turned to handling the manufacturing of their own surplus milk. This put the fluid milk network into competition with the cheese factory network. However, for fluid milk producers and processors, manufactured products were primarily a form of surplus control. As a result, the need to make a profit from the sales of these products was not as high as the need to simply get the milk off the fluid market. Consequently, they could easily underprice their products, basically dumping them on the market. The rise of dairy product manufacturing as a form of surplus control was economically devastating to the crossroads cheese factories and their farmer patrons. In addition, as the fluid milkshed expanded into their territories, cheese factories could not get enough milk to continue production. Therefore, the permeability of manufacturing and fluid milk markets was extremely problematic for both dairy production networks.

Inner-ring producers tended to belong to the Sheffield company–organized "captured" cooperative or other small "bargaining association" cooperatives, or they sold as "independent" (not organized) producers. The battles between Dairymen's League cooperatives and Sheffield Farms—often represented as populist versus company farmers—in fact represented conflicts between inner- and middle-ring producers.[45] Inner-ring dairy producers were hit particularly hard in the summer, when milk "flushed" out of the hills and outer areas with cheap land costs. The only recourse inner-ring farmers saw for themselves under these conditions was to try to exclude other farmers from the market. They did not see the promotion of a statewide cooperative as being in their interest.

Looking back, one can clearly see why leaders at the time, and historians who have written about this period, have emphasized the problem of unity. Farmers, organized collectively, could have negotiated prices with dealers. Then dealers could not have multiple-sourced various farmers off each other in order to buy milk at the lowest price. Yet in the previous period, when the dairy industry was separated into two markets, unity was not an issue. The political problem of collective organization and action became an issue only with the rise of fluid milk markets, particularly once the barriers between the two dairy markets broke down. In fact, part of the political problem of unity resulted specifically from the transition from a segmented to a unified market and the spatial restructuring involved in that transition.

For those in the middle ring of producers, market realities became increasingly uncertain in the 1930s. This was particularly true for those farmers who were situated close to railway lines yet far from urban areas, like the farmers of Heuvelton. Previously, the milkshed had extended for more than four hundred miles along these rail lines, taking in even distant counties like St. Lawrence and Jefferson Counties. Many of the farmers near railways in these counties were essentially in the inner ring of the milkshed until the arrival of truck transport. Like many inner-ring farmers, they belonged to the Sheffield company cooperative. These farmers had already made the investments to participate in the fluid milk market, yet they now found themselves pushed out of the market. When milk demand dropped during the Depression, farmers could not do away with the fixed investment, in better land, better machinery, and higher-producing cows, necessary to participate in that market. Fluctuations in demand increased this instability and uncertainty: rising demand in the city brought back opportunities to participate in the higher-priced fluid milk market, while contraction in demand contracted the milkshed and often left these farmers back in the manufacturing market.

Because they had the most to lose with the border instabilities of the 1920s and 1930s, farmers facing market instability in the middle ring were those who were the most radical in their fight for higher milk prices. When the north country farmers of St. Lawrence and Jefferson Counties were moved out of the inner ring, they left the Sheffield cooperative and joined the more radical Dairy Farmers Union. While information on membership in radical dairy farm organizations during the 1930s is limited, table 8.1 shows the county residence of Dairy Farmers Union members and convention delegates in 1943. This table shows that most members were from counties, like Jefferson and St. Lawrence, that had traditionally been part of the fluid milk market at the time of the milk train system. The table shows that farmers from counties with secure access to fluid milk markets did not tend to belong to this organization.

Table 8.1
Dairy Farmers Union Membership and Convention Delegates, by County, New York, 1943

County	% of Membership	Convention Delegates
Albany	1	0
Broome	0	0
Chenango	3	10
Clinton	1	0
Columbia	0	1
Cortland	4	3
Delaware	0	0
Essex	0	0
Franklin	9	8
Fulton	0	0
Greene	0	0
Herkimer	1	6
Jefferson	19	9
Lewis	3	0
Madison	3	4
Montgomery	1	0
Oneida	3	1
Onondaga	0	1
Oswego	4	6
Otsego	4	4
Rensselaer	2	5
St. Lawrence	29	12
Saratoga	0	0
Schoharie	4	5
Sullivan	4	6
Tompkins	0	0
Ulster	2	4
Warren	0	0
Washington	6	7

Source: Farmers Union of the New York Milkshed Papers. Division of Rare and Manuscript Collections, Cornell University Library.

A description of the location of the 1933 strike makes the geography of New York dairy farm radicalism clear:

> The milk strike of August 1933, which encompassed twenty-seven counties in central New York, was intense and violent. . . . The August action against the state, the League, and other dealers, began near Utica, spread to northern Lewis County and then west to Rochester. Major disorders occurred in central Oneida County. The strikers dumped thousands and thousands of gallons of milk both their own and that of their recalcitrant neighbors-usually members of the Dairymen's League.[46]

The geography described in this passage is the region of middle-ring dairy farming. These farmers were caught between the potential of moderately good resources and moderately good market access, which tended to keep them in business, and fluctuations in their access to the high-priced fluid markets, which kept them always in a state of uncertainty. Poorer hill farmers were more likely to slip into rural poverty–style subsistence farming or to abandon their farms altogether if their financial condition took a turn for the worse. If they found themselves unable to remain profitable, inner-ring farmers were likely to sell their valuable suburban land to other farmers or to developers. Middle-ring farmers, with adequate resources and no suburbanites willing to pay high prices for their land, saw a reason to fight politically to stay in business.

While inner-ring farmers experienced greater losses in bad times and outer-ring farmers experienced more absolute attrition in numbers, middle-ring farmers experienced more ongoing instability and uncertainty in their enterprise, and it is this fact that made middle-ring farmers the most radical. The instability in middle-ring market areas was due to the ebb and flow of the fluid milkshed. As a result, it was the "middle" group that was most likely to strike. In a history of New York state farm women and the 1933 strike, Ford notes that the Dairy Farmers Union, the radical dairy group that called the strike, appealed "to less established farm families," her term for the middle group between "established" and "marginal" families. This group had more resources to ride out a strike and less to lose by their radical actions. On the other hand, "Pro-'Establishment' farm women, perhaps in Home Bureau or Dairymen's League auxiliary, opposed the strike for philosophical reasons. In addition, some women on economically marginal farms feared and opposed a milk strike because they dreaded losing everything."[47] While "Establishment" farm women were likely to be "philosophical," some security in access to markets, due to being "established" in farm size and interest group/cooperative membership, most likely affected their opposition to the strike. Struggling "marginal" farms, often small farms on hills, with poorer soil, were closer to outer-ring farmers in their political position. They did not have the resources to ride out a strike.

This combination of instability plus resource potential made middle-ring farmers into dedicated organizers, fighters, and strikers. Yet this organization could only go so far, because of the intrinsic political differences between inner- and outer-ring farmers. Like inner-ring farmers, middle-ring farmers had to deal with the price pressure resulting from the spring flush of milk just outside the milkshed. However, they also had to fight against inner-ring political attempts to restrict access to the fluid milk market.

CONCLUSION

Recent events could be interpreted as a truce in the border wars. Since 1995, both cooperatives and dairy firms have undertaken massive merger activities. As a result, the one major super-cooperative, Dairy Farmers of America, has twenty-five thousand producers in forty-five states and is responsible for nearly a quarter of national milk production.[48] Concomitantly, two milk firms—Dean Foods and Suiza—have participated in significant merger activity since 1997.

In 1999 Suiza purchased Southern Dairy Foods, the country's third largest processor, in a joint venture with Dairy Farmers of America.[49] It has purchased thirty regional dairy companies since 1993. Dean Foods has also become a national presence by purchasing regional dairy companies. Current estimates are that these two national companies now own more than 10 percent and maybe as high as 20 percent of the fluid milk market.[50] These trends would seem to represent a return to a corporatist form of economic governance, with both producers and processors organized in large-scale, powerful organizations.

However, despite these happenings, interregional competition continues. With the increase in ability to transport milk long distances, particularly with new high-temperature pasteurization techniques, the regional scale of sectional dairy politics now encompasses the entire United States. The most visible form of regional market protection is the Northeast Dairy Compact, an agreement between New England states to give farmers serving their market higher milk prices. Midwest congressional representatives, whose dairy constituents are interested in entering northeastern markets, assail the compact as a restriction on trade.[51] New England supporters, on the other hand, describe the compact as a way to save family dairy farms in the region.[52] New York, while officially invited to join the compact, is not a member. As usual, representatives of the state's strong urban political constituency voted against joining the compact, because "cities fear the Compact would make milk more expensive, and of course it has."[53]

Throughout more than a century of dairy politics, the idea of perfection has provided the framework for the transformation of the dairy economy. The hegemony of milk as perfect created the vision of perfect production, which separated farmers into those that were modernizable and those without the natural, economic, or locational resources to meet the ideal. This posed an obstacle to political unity among dairy farmers. Second, the idea of the perfect milkshed, as a law-like determination of the spatial rings of dairy production around cities, denied the political border struggles that have been such a salient feature of dairy markets for many decades.

Once again, the perfect story became an act of power, which separated out ideal and less than ideal forms of production. Yet this story tends to produce "border-line problems," since the boundary separating the perfect from the imperfect ultimately becomes a site of political contention.[54] How dairy states responded to this political problem is the subject of the next chapter.

9

Alternative Visions of Dairying

Productivism and Producerism in New York, Wisconsin, and California

THE "SOCIAL" IN the social embeddedness of markets becomes particularly clear if we focus on the setting of market boundaries—in particular how certain participants are included or excluded from "entry" into particular markets. Comparing New York's milkshed governance to a very different form of boundary politics in Wisconsin and California will show that these states operated in significantly different political cultures. New York's policies, based on the consumerist political culture of the industrial bargain, emphasized "productivist" solutions: raising farmer income by raising agricultural productivity. Because the political culture of Wisconsin and California included more powerful agricultural interests and weaker urban constituencies, the industrial productivist vision carried much less force in these states. Instead, agricultural experts in these states followed a more "producerist" policy that emphasized farmer political and economic organization as part of the solution. Producerist policies tended to emphasize the diversity of farm strategies depending on a farmer's resources and the institutional linkages he or she forged beyond the farm gate. In other words, the impersonal, law-like effects predicted by von Thünen's spatial models were mediated by local, contingent political action in determining how city milksheds would be governed.

Fluid milk companies, for their part, practiced the strategy of multiple sourcing in milksheds across the country, leading to increased farmer sectionalism and violence, and to market disruption and crisis. The state, and eventually federal, response was a set of New Deal laws that allowed for the establishment of relatively stable market boundaries in the dairy industry. Until the establishment of market orders, the interstate commerce laws had made most restrictions of markets illegal. The agitation in favor of legalizing control of production by agricultural cooperatives, eventually passed as the Capper-Volstead Act, along with the establishment of market order legislation gave states more power to control milkshed borders.[1] However, the use of these and other powers varied by state, according to the nature of sectional conflict within milksheds.[2] The institution of market orders gave farmers the right to establish formal market rules and restrictions.[3] The formal, institutionalized setting of market boundaries is one aspect of the design of market order "governance": the way state institutions structure market competition.[4] Yet, in producerist states, farmers were able to use federal market order law to frustrate the multiple-sourcing strategies of dairy companies. Productivist states with a strong consumer politics, like New York, used the federal market order system in a much weaker way, enabling companies to continue multiple-sourcing strategies to a significant extent. The major difference in policy involved the extent to which states set strong boundaries between fluid milk and manufacturing milk producers. The three major dairy states—New York, California, and Wisconsin—all experienced periods of dairy industry instability and violence. Yet responses to the boundary crisis in the three states differed substantially, despite the fact that each state used the same federal market governance laws to resolve the crisis.

The overarching term commonly used to represent the politics behind agrarian social movements is "populism." In its simplest terms, populist agrarian movements are defined by their belief that large agribusiness monopolies—and their large farmer partners—are corrupt and that ordinary farmers need to get together to take back political power. Yet only in producerist states with powerful farmer political organization could populism develop as a significant political movement.[5]

In producerist states, populist agrarian movements developed beyond the simple antimonopoly stance to formulate alternative economic visions for the nation. From the Greenbacker, Granger, and Free Silver movements of the late nineteenth century to the Farmers' Holiday, Dairy Farmers Union, Society of Equity, and Nonpartisan League of the twentieth century, populist farm leaders presented alternative economic visions that challenged the industrial ideal of perfection. These visions went beyond simply advocating a diversity of agricultural strategies. Instead, alternative visions attempted to create new forms of collective action that challenged productivism as perfection.[6]

Social histories of New York state dairy politics tend to place milk strikes and other forms of agrarian resistance into this populist framework.[7] Yet in New York state, dairy agrarian protest did not actively challenge the ideal of perfect production, but centered primarily on milk price and sectionalist competition for access to this higher-priced market. The industrial vision of dairying and its productivist ideology became the solution to farmers' problems.

Why was New York's dairy policy, unlike that in Wisconsin and California, so dominated by the industrial ideal of perfect farming? The answer to this question has to do with the politics of the industrial bargain. Land grant colleges survive on funds provided by state legislatures. Because of New York state's urban consumer-dominated demographics, politicians in that state garnered votes based on their ability to serve urban consumers. In other words, because consumer interests in the state were so powerful, the industrial bargain was particularly binding, forcing the state's agricultural institutions into a pro-industrial, productivist form of politics. Experts at the state's "land grant" agricultural college at Cornell University, along with public health officials, supported productivist policies—the encouragement of a large-scale, efficient, highly productive form of dairy farming—in order to provide urban consumers with safe milk cheaply.

Elizabeth Sanders's work comparing farmer-labor politics in the Northeast, Midwest, and West can help explain the political differences between these regions. She argues that agricultural politics in the late eighteenth and early nineteenth centuries was simply more important in states where it was the major industry.

> There were, and are, agricultural areas within the manufacturing-belt states, but they are dependent on the industrial cities for their raison d'etre and not the reverse. The dairy, truck and poultry farms outside the industrial cities existed to supply the city residents with foodstuffs, particularly perishables. In the agriculture/extraction-based periphery [of the West and Midwest], on the other hand, the engine of economic growth lay in the countryside.[8]

As a result, western and especially midwestern farmers were able to forge coalitions with labor groups and to elect leaders more sympathetic to both interests. In "manufacturing-belt" states, like New York, labor groups were more likely to ally with urban consumers.[9] These differences in political culture meant that state officials in Wisconsin and California were not as committed to the industrial bargain between consumers, large-scale processors, and larger farmers. Instead, these states were able to implement policies that differed from the productivist ideology of the industrial vision.

New York, Wisconsin, and California used a number of tools to manage the spatial structure of dairying within their territories. While federal market orders set up a method of legally reestablishing the spatial boundary that was previously physical, states for many decades had significant control over the permeability of this boundary. New York, Wisconsin, and California used health regulations, cooperative organization, and market orders to manage the spatial structure within their territories.

However, New York was never able to set strong boundaries between dairy networks. As noted earlier, there were two ways the boundary could be set. The first was to make a tighter, more exclusive boundary in which a smaller number of farmers were able to sell most of their milk into the fluid milk market and thereby receive a high "blend price" because little of their milk went to manufacturing uses. The second way was a more inclusive policy in which more farmers in a wider milkshed shared a lower blend price.

The political pressures constituting the industrial bargain encouraged a wide market boundary. In many upstate counties, dairying was the primary economy, and political constituencies in these counties depended on politicians to maintain their market access. For urban politicians, on the other hand, cheap milk was the important issue, and a wider milkshed boundary created enough price competition between farmers to keep prices low. Politics in New York state has historically been split between upstate Republicans, representing farmers, rural people, and smaller industrial cities, and more numerous urban Democrats, representing downstate, urban interests.[10] A pro-industrial dairy policy yielded higher incomes for upstate Republican farm constituencies and cheaper milk for downstate urban Democratic constituencies. Smaller, low-resource farmers were left out of this productivist/consumerist alliance.

New York's milkshed boundary governance, therefore, set strong boundaries at state lines, but was more inclusive within the state. The next section will describe how New York state policy makers used various tools to implement this productivist form of dairy market regulation. The subsequent sections will describe the more producerist implementation of dairy market regulation in Wisconsin and California.

BOUNDARY GOVERNANCE TOOLS AND THEIR USE IN NEW YORK

Health Regulations and Licenses

As the boundary crisis intensified, officials turned to health regulations as an informal way to form market boundaries around milksheds.

> At the time of the Great Depression, local sanitary regulations were widely used to erect barriers to the movement and marketing of milk. These included refusal to inspect farms of plants beyond the existing milkshed, outright exclusion by ordinance of milk from plants outside the city limits or a prescribed radius, excessive fees for inspection at a distance from the city, and many others.[11]

One reason cities limited inspection was, as noted earlier, their limited bureaucratic capacities. Budgets for inspectors were small, and therefore to fully cover the milkshed with a few inspectors required restricting the number of farms and firms. New York City restricted the number of milk-selling firms by setting up a licensing system for dealers. Dealers without a license were not able to sell their milk in the city. The advent of the truck transportation system for milk intensified the efforts to use health inspections as a way to control milkshed boundaries. However, after decades of legal challenges, courts officially stated what officials informally noted all along: New York state tended to favor in-state firms and their associated farms.[12]

A map of the New York milkshed in 1925 shows the extent to which New York City went north, rather than east or west, to get its milk (fig. 9.1). The milkshed only reached into the border counties of the large milk-producing state of Pennsylvania. These other states served other northeastern cities, but there were still plenty of farmers in those states without access to fluid milk markets.

Over the decades, these barriers were legally contested and overturned in New York.[13] Yet these policies were in place for nearly a

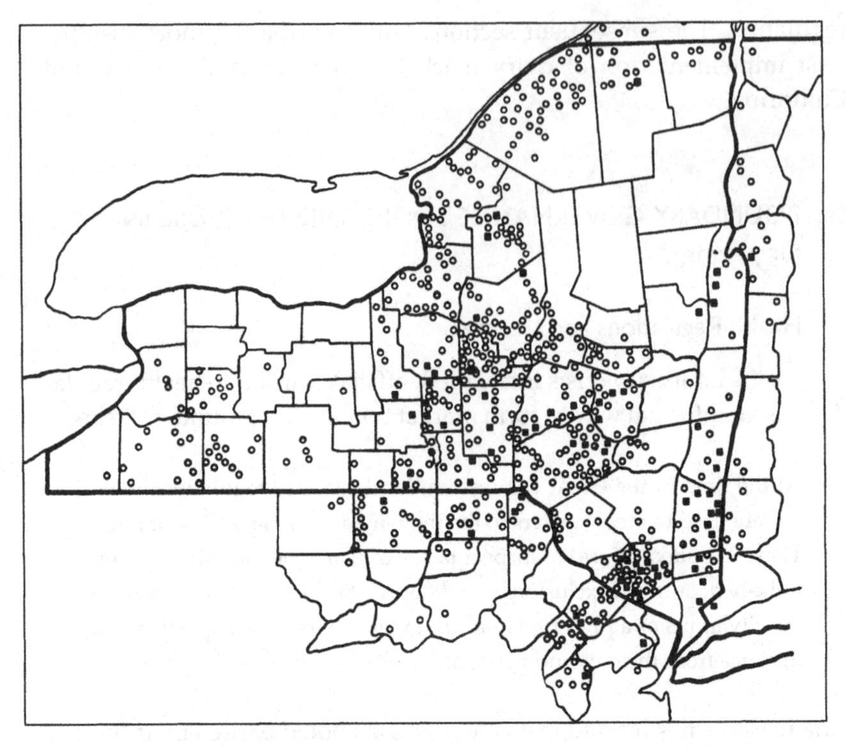

FIG. 9.1. New York City milkshed in 1925. From Norton and Spencer, *A Preliminary Survey of Milk Marketing in New York.*

century and have left their legacy in terms of the spatial organization of the state's milkshed.

Cooperative Policy

The permeability of the milkshed was primarily managed through the state's policy toward dairy cooperatives. New York state agricultural officials have historically supported a state-wide cooperative with open membership: the Dairymen's League. In the 1930s, state officials encouraged the cooperative to sign up as large a percentage of farmers as possible, thereby encouraging the cooperative to reach into all dairy areas, including the summer dairy hill farm areas.[14] The Dairymen's League, for its part, believed it could gain control of milk prices through the management of surplus milk in its league-owned manufacturing plants and through its contract to supply milk to Borden's Milk Company. Yet this overexpanded system soon became untenable, since it brought down the price cooperative members received—due to a low blend price and payments for investments in manufacturing plants. As a result, farmers with other fluid market alternatives left the league, leading to serious financial difficulties for the cooperative.[15]

As noted earlier, inner-ring milk producers did not have the political power in New York to set a strong boundary between themselves and their middle- and outer-ring counterparts. The early extension of the milkshed into the outer territories during the milk train system made it difficult to pull the market away from these areas when milksheds restructured around tank truck transportation. State policies that tended to support farmers' association with the Dairymen's League also made it difficult to create strong boundaries between fluid and manufacturing milk producers. In particular, state influence to make the Dairymen's League the representative of all dairy farms in the milkshed created conditions where more farmers felt they had the right of access to fluid markets. However, the sectionalist conflict between farmers made it difficult to establish a unified, statewide cooperative. This made the Dairymen's League politically and financially weak.

Market Orders

The formal boundary governance solution—initiated at the local state level and eventually instituted as a federal policy—was milk market order legislation.[16] The Agricultural Marketing Agreement Act of 1937 enabled states to implement marketing orders, a set of rules that states could use to create stable markets in their territories. Milk market orders stabilized dairy markets by establishing standardized pricing for fluid milk. Fluid processing plants that served urban markets were part of these orders, while manufacturing milk plants were excluded. Market orders therefore established the boundary between one market, for fluid milk, and another market, for manufacturing milk.

Market orders formally instituted a pricing system that had existed between market milk farmers and their milk buyers for many decades:

> The added cost of producing inspected milk has been used to justify a higher price under classified price plans for milk used in packaged

> fluid milk products than for milk used in manufactured dairy products. Believing that low prices might reduce the supply or force farmers to relax standards in production, legislators have justified public regulation of milk prices partly by the need to assure the public an adequate supply of wholesome milk.[17]

The milk market order created provisions for setting a legal boundary between intensive and extensive dairy systems, by making processing plants that served city milk markets members of an order pricing pool, while leaving manufacturing milk plants outside this pool, and therefore outside the pricing system. Market orders reestablished—by law rather than by the location of milk train routes—the spatial segmentation of the dairy system. This legal system created a relatively "closed" economy through the creation of a protective market boundary.

The institution of this boundary helped to return stability to the system. Once farmers knew what prices they could expect to be paid for their milk, they could establish the intensiveness of their dairy operations according to reasonable cost and return expectations.

But even with the order system in place, the permeability of boundaries differed from one milkshed to the next.[18] While the mechanism for the creation of a milk market order was federal law, management of the boundary between order plants and non-order plants was primarily the responsibility of individual states. Surplus milk is the milk that is sent by farmers to market order plants but that does not end up in fluid use. Like milk produced outside the market order boundary, this milk goes into the production of manufactured products that can be stored. For many decades, states treated the pricing of surplus milk differently, a difference that reflects the relative permeability of the boundary between the two dairy economies.

New York chose the more permeable, more inclusive pricing systems available under the order: a "market-wide" pricing system in which all farmers in the milkshed shared the same "blend price." This system worked in much the same way as the Dairymen's League pricing had since the 1920s, yet it was now implemented across the milkshed (although league members still had the extra burden of paying for their manufacturing plant costs).

Land Utilization Planning

In 1926 the New York State Commission on Housing and Regional Planning released a regional plan for New York state, the first statewide regional plan in the United States.[19] This plan advocated direct state involvement in the "planned abandonment" of what planners referred to as "marginal or submarginal" rural land.[20] The plan recommended that the state take upon itself the role of determining which land was suitable for permanent cultivation and which land should be taken out of cultivation, purchased with public funds, and reforested. The goal was to "rate the land according to its agricultural possibilities, to classify the roads in its vicinity and to determine where electrical lines should be located to serve the farms remaining in these areas."[21] Governor Roosevelt and the New York state legislature put the purchasing power of the state behind an effort to make the classification of uses recommended in the planners' maps a reality. The Hewitt amendment appropriated over $1 million, and scheduled the appropriation of more funds for the purchase of land classified as "submarginal."[22]

Increasingly, planners talked about less intensive hill farms as a form of social pathology. For example, as president, Roosevelt set up a National Resources Planning Board. One of the major pieces of work this board produced was a lengthy and influential report on rural land problems. The report argued that "the decline of rural civilization" was due to "pathological modes of [land] use."[23] This "maladjustment" in land use had led to a "socially degraded existence" for the occupants of these marginal lands "as a result of their inadequate income, poor schools and roads, and infrequent contacts with outside civilization."[24] The solution, according to these New Deal planners, was to "rehabilitate" families by moving them off these poor lands.[25]

The idea that "the land problem" was a social pathology—and land use planning the cure—was repeated frequently in public documents on the issue during this period. For example, one article on the benefits of land utilization studies stated that submarginal areas "make government expensive and inefficient, encouraging poor schools, bad roads, remote and feeble government, and low levels of community life."[26] If land use were planned through land classification studies, "areas of sparse settlement, and low taxable wealth values would not be allowed."[27] Government land use planning officials made the diagnosis that the relationship between humans and the environment was diseased and that the government, as doctor, needed to provide the cure.

A 1935 New York State Land Planning Board report presented the state as both healer and protector. Entitled *Submarginal Farm Lands in New York State,* the report portrayed farmers in poorer soil areas as victims of shady land speculators. According to the report, most of the residents of these farms were people who had been tricked into buying land that could not produce and who were therefore trapped in lives of grim poverty because they could not sell their land. Selling the farm to another, however, did not solve the problem, because it left "another family marooned on the island of want."[28] By purchasing the land, the state would both "rehabilitate" people who had made this mistake and protect other people from repeating this process. Like the national report, the New York state report recommended the "rehabilitation" of these families through resettlement.

These reports seldom mention farm practices. Rather, the New York report asserts that "large areas of the state are not contributing adequately to its welfare," thereby focusing blame on the land itself rather than on the farmer.[29] Space itself had become the "noncommunitarian" factor, and the farmer simply a passive victim of the land. In fact, marginal farmers in this report are seldom referred to as farmers at all, but as "persons who did reside in the poorer agricultural lands."[30]

According to these reports and speeches, the pathology exhibited in these cases was an inability to obtain a "modern" standard of living. To New York state land planners, a marginal farm was an "island," a remote and inaccessible place where people were trapped away from the "mainland" of modern life. Therefore, the only way to rehabilitate these people was to remove them from the land. One discussion of the land utilization program defined "submarginal land" as "land low in productivity, or otherwise ill-suited for farm crops, which falls below the margin of profitable private cultivation."[31] Planners, in other words, defined nature as marginal to a particular agricultural economy if it did not yield profits equivalent to those of an intensive farm. The margin in this case was the economic margin of income resulting from the use of productive inputs.

Sociologists also had begun to use the term "marginal" in their work, to describe people left behind by, or unable to cope with, progress.[32] It was a short step from the argument that farmers on low-resource lands were economically marginal to the argument that these farmers were socially marginal as well. In fact, discussions of the concept of marginal land and of marginal farmers tended to link the economic with the social definition of marginality. Lane's description of marginal farmers living on an isolated island and Gee's description of "sparseness of settlement" as unhealthy echoed the prevailing social viewpoint that this isolation was an obstacle to modern development.

From an "ecological" viewpoint, New Deal planners argued, correctly, that some land was never meant to be farmed—that nature determined to a great extent how land could be used. However, this ecological rhetoric also legitimized policies that deemed only one form of agricultural production ecologically adaptive, namely, the intensive system, and deemed other forms of agricultural land use pathological. Land use planning policies, based on this viewpoint, worked primarily to determine which land could be agriculturally "modern"—that is, intensely cultivated—and which land was incapable of achieving this status. The policy solution was the eradication of other forms of farming—and living—through planning. Implementation either involved either planning infrastructure improvements to avoid these areas, identifying these areas for banks and encouraging them to redline loans to these areas, or, in some cases, using national Resettlement Administration funds to remove people from these areas.[33]

The classification method focused on soil class but also considered factors like the market access and condition of farm buildings.[34] The land use classification system mapped where intensive dairy production was possible. Agricultural economists mapped the entire landscape of New York state, dividing land "into five classes ranging from land earmarked for public ownership and early reforestation to land rated for permanent retention in agriculture."[35] While other states did land classification studies, New York began them earliest (the 1920s), continued them longer (into the 1960s), and carried them out most intensively.

Land use planning policy documents do not address the implications of spatial restructuring on alternative forms of dairy land use. Instead, conservation—specifically reforestation for recreation, soil conservation, watershed improvement, and wildlife preservation—was continually cited as the major rationale for land use planning in New York state. Yet extension policies at the time explicitly included a "statewide campaign for the removal from the market of the unfair and unnecessary competition of milk produced at a loss."[36] While land use planning policies may not have been enacted specifically to restructure the dairy industry, in fact, affected interest groups were not unaware of the implications of conservation policies on the structure of dairy agriculture in New York state.

For example, Cornell worked closely with the New York State Farm Bureau's Committee on Land Utilization, Roads and Electrification as well as with the Farm Conference Board's Farm Light and Power Committee, which advised Cornell and Albany officials on rural land use planning. Members of the Farm Conference Board, including representatives from the Farm Bureau, the Dairymen's League, and the Grange, supported the idea that only land in the permanent farm class was to be the focus of better infrastructure development.[37]

These two committees continually reminded utility companies and the Public Service Commission of the differences between those farm areas that were classified as permanent and those that were not. The Farm Light and Power Committee organized country committees "for the purpose of contacting the utility companies and laying out and approving proposed line extensions."[38] The Farm Bureau also had a great deal of say in the appointment of Maurice Chase Burrit, a Cornell agricultural economist and a former director of Cornell Cooperative Extension, as a commissioner of the Public Service Commission, which was in charge of approving lines for rural electrification.[39] Burrit had drafted the Cornell Cooperative Extension policy statement that recommended a campaign against "milk produced at a loss." At the commission, Burrit was a strong proponent of land planning through control of rural electrification, as letters between Burrit and Farm Light and Power Committee officials make clear. In a letter dated June 3, 1937, Farm Light and Power Committee general secretary Edward S. Foster wrote to Burrit concerning his recent address to the Telephone Association:

> I think your suggestion that the Telephone Companies give more thought to land utilization surveys in developing their lines is particularly fitting. It is our belief that every farm in the permanent farming areas of the state must eventually have access to good roads, good schools, electric power, telephone service, and R.F.D.[40]

When farmers in the hill farm areas began to demand access to infrastructure services as well, in a letter dated March 13, 1940, Foster suggested that Burrit talk to utility officials:

> If you get an opportunity to do so I think it would be well to talk with Mr. Buckman and Mr. Kelsey with reference to questions which have been raised in Chenango County. I should dislike very much to have these extension plans jeopardized by any attempt to counsel electrification of all farms in Classes 1 and 2 [the low-quality soil land classes].[41]

Reforestation programs in New York state were therefore part of a struggle by urban, industrial, and state planning elites to impose an industrial vision of production on the state's agricultural resources. New York rural land planners actively rejected the idea of two dairy systems in the state and consequently favored a policy of "publicly directed abandonment." As the New York State Land Planning Board report states,

> Seldom is it profitable to engage in a type of farming that differs radically from that in neighboring areas. Those areas classified as submarginal cannot compete either with the regions where radically different types of farming prevail, or with the fertile and productive farming areas right in their own neighborhood.[42]

New York state land planning policy was also unusual in that decisions were carried out by a small number of officials. In contrast, land planning in Wisconsin involved institutionalized public input and participation by the farmers and other local people, as described in the next section. New York state planning professionals, in contrast, labeled low-resource pasture-based farm systems first "submarginal" usurpers of public resources and later victims of land use "pathologies" against nature. In accordance with powerful New York state urban constituencies, these planners advocated public forests as a form of moral revitalization through recreation.

Land utilization policies were carried out primarily in the name of soil conservation. The pathological mode of land use on hill farms was therefore imperfect not only due to its low productivity, it was also considered a form of ecological degradation. Yet making the intensive, industrial farm the perfect form of dairying has since resulted in a new set of environmental problems, such as water pollution due to manure, fertilizer, and pesticide runoff.[43] Therefore, while these policies were enacted in the name of conserving nature, in fact they encouraged a form of agriculture that has become increasingly unsustainable.

WISCONSIN: POPULISTS CHALLENGE PRODUCTIVIST IDEOLOGIES

States that did not have powerful urban constituencies were more likely to see their political interests linked to farm interests.[44] As a result, Wisconsin dairy farmers were able to maintain a more populist form of political culture that could publicly challenge the ideal of industrial dairy production. In Wisconsin dairy farmers successfully carried out a producerist politics and prevented the growth of a productivist ideology of its land grant institution in Madison.

The role of Wisconsin dairy farmers in the election of the populist governor Robert La Follette illustrates their political power in the state. According to the historians Saloutos and Hicks, La Follette

> found his hopes for political advancement blocked at every turn by a party machine subservient to the state's industrial leaders. Taking his case to the people, he persuaded a majority of them, farmers for the most part rather than city dwellers, to back him in his war on the bosses.[45]

As described in chapter 4, New York City's early demographic growth had enabled a vocal urban consumer reform community to gain control over the state's agricultural policies. In contrast, Wisconsin's less urban demographic profile meant that dairy policy was less open to urban consumerist political influence. As a result, rural interests had a greater and more sustained voice in state agricultural policy.

In addition, smaller Wisconsin farmers were numerous and better organized. All Wisconsin farmers, both large-scale and small-scale, have had a strong political voice in state agricultural policy, each through their own political organizations. As in other states, the Farm Bureau promoted scale, efficiency, and productivity and therefore represented larger farmers with the resources to implement such on-farm strategies.[46] The Wisconsin Society of Equity forcefully represented smaller farmers.

Wisconsin farmers as a whole were a more unified political group, for a number of reasons. First, over three-quarters of the state's dairy farmers were of German or Scandinavian ethnicity, and this similarity of background assisted in maintaining their social solidarity.[47] Second, the political geography of Wisconsin's dairy industry encouraged alliances and discouraged sectionalism. Three-quarters of Wisconsin's dairy output was sold out of state.[48] Wisconsin middle-ring and outer-ring farmers could therefore fight together against a common enemy: all other states restricting their access to out-of-state milk markets. Unlike New York or other urban dairy states, Wisconsin saw little competition between consumers and producers. This fact enabled Wisconsin dairy farmers, and their state leaders, to attain a unity of purpose in legislation and in cooperative organization not possible in New York, where the urban market was also a state constituency. In Wisconsin the power of local dairy cooperatives and the relative weakness of the Farm Bureau helped to create the conditions for this political unity.

Combined with the out-of-state markets was the fact that the state as a whole was dependent on the dairy economy. Many Wisconsin regions depended heavily, almost exclusively, on dairying as a source of regional income. In New York the dairy manufacturing sector was just one small part of a significant upstate manufacturing economy. In contrast, one study of a Wisconsin town describes efforts to build a local industry: "By dint of much effort, weekly pay rolls in this center were built up during one period, including those employed in milk plants, to about $4,000. This proved to be less than one-sixth of the amount paid each week by the two dairy plants, to their farmer patrons."[49]

Cheese plants dotted the Wisconsin landscape long after small cheese factories disappeared from New York's crossroads. They represented the major rural manufacturing sector. Because they were faced with few economic alternatives, Wisconsin state officials and academics paid more attention to the needs of small cooperatives, small processors, and the more extensive type of farmer, who represented an important part of the rural economy. In addition, many of the local cheese plants were still cooperatively owned, as opposed to small New York cheese plants.[50] Producerist politics represented all of these economic interests.

Most Wisconsin fluid milk producers were middle-ring producers for the Chicago market, and cooperative membership among this group has historically been strong.[51] Wisconsin, however, unlike New York, has historically advocated small, local cooperatives.[52] The one attempt to form a statewide cooperative managed to garner only 8 percent of the state's dairy farms and received little support from state officials.[53] Unlike officials in New York, Wisconsin officials rejected both market-wide cooperatives and coordination with the Farm Bureau as a solution to dairy problems. "Wisconsin never adopted the 'Farm Bureau' plan of setting up responsible farm groups within each county to direct the public Extension program," states one cooperative extension history.[54] Instead, the state set up "township committees" to work with and advise county agents. As a result, the Farm Bureau had less control over dairy politics in Wisconsin, particularly compared to its strength in both New York and California. Wisconsin's agricultural cooperative extension service "willingly cooperated with private or quasiprivate organizations like the banks, railways, and mercantile agencies. But Wisconsin firmly prevented the newly formed Farm Bureaus from dictating policies to the Extension forces or permitting the county agent to become a membership solicitor for farm bureaus."[55] Instead, the Wisconsin Society of Equity represented "[t]he strongest and most vocal of the farm organizations with whom Wisconsin Extension workers cooperated."[56] This actively populist group vocally represented the interests of smaller farmers.[57] The Society of Equity criticized the industrial vision of production being promoted in the state's major agricultural institution, the University of Wisconsin at Madison. Instead, it pushed university extension policy and research toward the producerist goals of marketing and cooperative development as a way to increase farmers' income.[58] State legislation mandated that the state agriculture school support cooperative marketing in the state.[59] As in New York and California, the larger farmers in the Wisconsin Farm Bureau believed that the role of the land grant college was to increase a farmer's income by increasing his productivity.[60] The Wisconsin Society of Equity, however, prevented the close Farm Bureau–land grant relationship that existed in many other states, including New York and California.[61]

Because populist agrarian interests have generally exercised considerable political power in the state, Wisconsin dairy policy has remained populist and pro-producer: produc*erist* rather than produc*tiv*ist. Producerist politics tends to represent farmers not simply as cheap food providers for urban constituencies but as vital and necessary contributors to the communities in what is a mostly rural state. In other words, producerist ideology, represented historically by the milkmaid, puts the farmer back into the picture, next to the cow. So long forgotten in most city-country politics, this ideology still retains some meaning in Wisconsin and some other midwestern agricultural states. As a result, Wisconsin's form of dairy politics has always differed substantially from New York's.

In New York, pro-consumer politics has caused Cornell's experts and the state's cooperative agricultural extension system to concentrate expertise and attention on productivist goals: creating larger, more intensive farms and ignoring or actively discouraging small farms. In Wisconsin, in contrast, small was not automatically translated into inefficient and marginal.

As a result, agricultural economists at the University of Wisconsin were ideologically opposed to a statewide cooperative, as reflected in the many invectives by university officials against Aaron Sapiro, a lawyer who was the leading advocate for large, region-wide cooperatives in the 1920s and 1930s. For example, the agricultural economics professor Marvin Schaars reminisced in an oral history interview, "there would hardly be a meeting with that class when Mr. Macklin did not take a jibe at Aaron Sapiro. He disagreed violently with Mr. Sapiro's plan of organization."[62]

Instead, the state encouraged local, small cooperatives. This small cooperative policy meant that the organizational boundaries between manufacturing and fluid milk production remained strong. Fluid milk cooperatives were better able to restrict membership to the local area—with the help of sanitary regulations—thereby giving their members a high blend price in the market order. Madison city regulations enforced market boundaries so closely they were eventually challenged and held unconstitutional by the U.S. Supreme Court.[63] Fluid milk cooperatives also demanded "full supply contracts" from milk companies, an agreement in which the company purchases all milk supplies from the cooperative.[64] The cooperative, in turn, supplies the company with only the amount of milk it needs. This type of contract strongly inhibits the use of multiple-sourcing strategies and, in cases where cooperatives control most of the milk in a region, reflects the strength of producer organization. This puts a cooperative in a powerful position in the market.[65]

Fluid milk cooperatives in Wisconsin have also had more influence over their own members. Wisconsin and other midwestern cooperatives have been in a better position to control market access to farmers and company access to milk supply. For example, some Wisconsin cooperatives have been able to differentiate prices to various producers within the cooperative by the amount of participation of members in the fluid milk market.[66] This member price discrimination ability has reflected the lack of alternative access to the higher-priced fluid milk market in the state. For example, in 1960 cooperatives handled 99 percent of the milk in the Minneapolis-St. Paul market order. Exclusivity of milk control has increased with the mergers and federations of dairy cooperatives in the last three decades. Even those cooperatives not under federation coordinate activities.[67]

As mentioned earlier, New York's market orders priced milk to farmers according to a "market-wide" plan. The Wisconsin order, on the other hand, priced milk to farmers according to a "base-surplus" plan for many decades. A "base" is an amount that a producer has legally allotted to the fluid milk market, usually according to the producer's history of production for this market. Agricultural economists have argued that base-surplus plans restrict entry into order markets more strongly than other forms of producer pricing.[68] The establishment of a base for new participants in the market could be difficult, since the base, in fact, was calculated according to one's historical participation in the fluid market. New entrants were allowed into the pool only when officials perceived a need to expand the market beyond the current producers. Historically, this was usually accomplished through expanding the network of health inspections required for Class A milk and through cooperative membership. Eventually, legal challenges successfully argued that these plans constituted restrictions of trade, and no order uses a base-surplus plan today.[69] Yet for many decades, the establishment of a base was a legal hoop that farmers in many milksheds had to jump through in order to gain access to the fluid milk market. Wisconsin also did not use market-wide pooling for many decades. Instead, each cooperative formed its individual pool, enabling cooperatives with more access to fluid markets to give higher blend prices to their members.

In Wisconsin, unlike New York, there was no possibility for most farmers to attain access to the fluid milk market. There was also no need for state officials' allegiance to larger companies, since there was no in-state urban populace lobbying for safe, year-round, inexpensive fluid milk. As a result, the industrial bargain was a weaker force in Wisconsin. Manufacturing milk farmers were the most numerous, the most unified, and the most politically influential, due to their enormous contribution to the total state income. In addition, Wisconsin was able to establish a strong boundary between manufacturing and fluid milk markets, primarily because the percentage of milk going to fluid markets has always been small and managed by cooperative membership.

LAND UTILIZATION PLANNING

Like New York, Wisconsin carried out an intensive rural land planning program that covered the state. Wisconsin also initiated the concept of rural zoning. Wisconsin's approach to land utilization planning, like New York's, was intended to facilitate a process of what was seen as inevitable land abandonment. Nevertheless, Wisconsin carried out this process in a significantly different way from New York state. Wisconsin's agricultural cooperative extension service established land use planning committees at the town and county levels. These committees were composed of local citizens, farmers, and cooperative extension agents.[70] As the cooperative extension handbook for this process stated, "*What is desired most in county agricultural planning is the carefully formulated opinions of the people themselves,* based upon *factual* information as will aid them in understanding their land use problems and the measures which may be used in their solution."[71] These committees, working with state-set guidelines, were responsible for deciding which land should remain in agriculture and which should be zoned nonagricultural. In addition, land classification by Wisconsin county committees concentrated primarily on land that was tax delinquent at the time, rather than on economists' predictions of whether particular areas had the resources to remain in farming in the future.[72] In contrast, in New York, a few agricultural economists and the Farm Bureau/Public Service Commission committee were responsible for determining which land held agricultural potential and which did not.

CALIFORNIA: PROTECTING REGIONAL DIVERSITY

Southern California dry-lot farms more closely fit the industrial mold, with their large herds, purchased feed and forage, full-time employees, and farmer managers.[73] Yet California, despite its long history supporting industrial agriculture, has historically maintained strong boundaries between market milk and manufacturing milk areas, and its dairy policies have tended to preserve the integrity of regional dairy strategies.[74] While California, like New York, had a large urban populace, particularly by the end of World War II, it is also an enormous state with many areas not adapted to large-scale milkshed transportation systems due to high mountain ranges. As a result, the major dairy farm regions in the state tended to be isolated from each other. Some of these farm areas have traditionally practiced more extensive or mixed farming strategies.[75]

As in Wisconsin, farmers have been a strong political force in California through most of the state's history. However, California agricultural interests have differed substantially from Wisconsin's populist politics. In California, large-scale commercial agriculturalists had more influence than smaller farmers. The California Farm Bureau, the Associated Farmers, and the California Chamber of Commerce's Agricultural Department all worked on behalf of larger farm interests. While smaller farmers were organized in the California State Grange, they had less of a say over state agricultural policy. The result was a farmer-led productivism, a movement toward a more industrialized form of agricultural production largely managed by farmers themselves.[76]

In Los Angeles, inner-ring producers were able to exclude middle- and outer-ring producers so as to become exclusive providers of milk for the city markets. In contrast to New York's four hundred-mile milkshed, more than half of the milk for the city of Los Angeles was actually produced within fifty miles of the city until the 1960s.[77] These figures also have a great deal to do with the differences in the type of dairy farming that went on in New York and Los Angeles at the time. Los Angeles dry-lot dairying is very intensive, with larger herds eating a great deal of feed not grown on the farm. On the other hand, New York dairy farmers were, and are still today, primarily dependent on farm-grown feed that requires substantial farm acreage. However, the control inner-ring producers had over access to fluid markets in Los Angeles cannot be attributed to the type of farming or type of soil alone, as the discussion of California state market order policy below will show. Conversely, the incredibly broad access to fluid markets enjoyed by New York state producers historically was not simply due to the market "efficiencies" of this system. More than one report declared the size of the fluid milkshed in New York state inefficiently large.[78] Not surprisingly, the restricted fluid milkshed in Los Angeles was also criticized for its inefficiencies.[79] These inefficiencies were the result of the political power of California dairy farmers in the inner ring.

Unlike New York, which experienced its strongest market growth in the 1910s and 1920s, California experienced its strongest increase in market demand in the 1940s and 1950s, spurred on by growth in the Southern California region during the Second World War. As a consequence, the greatest competition for markets between inner-ring, peri-urban and middle-ring farmers—the type of competition that New York state experienced in the 1920s and 1930s—occurred in the 1950s in California. At that time, San Joaquin dairymen supplying "country plants" for the Los Angeles market formed a new producer cooperative. Through this cooperative, they attempted to gain control of all the fluid milk supply moving out of the San Joaquin region, thereby gaining the ability to control the price of this milk. In effect, these dairy farmers were attempting to form the type of market-wide cooperative long advocated by officials in New York state. However, these producers had one main problem: they did not control the inner ring. But because the cost of production for milk in the San Joaquin area was lower than inner-ring dry-lot dairying around Los Angeles at this time, these farmers were confident that milk processors would feel it was necessary to negotiate for their milk. Instead,

> Distributors evaded the cooperative, and supplier contracts were cancelled when producers would not withdraw membership [from the cooperative]. In their place new contracts were distributed to other farmers and contracts in the Southern milkshed were increased. Many vacant dairies in Kings, Tulare and Merced counties bear witness to the producer-distributor war, and new dairies were initiated by the same stimulus. Also, as a result, processing companies now distribute supply contracts cautiously in order to allow no one area to feel that it occupies a dictatorial position. . . . Had it not been for this abortive attempt by the bargaining cooperative, a greater proportion of Los Angeles milk supply might now have originated from the San Joaquin Valley.[80]

In response, inner-ring dairymen around Los Angeles organized themselves into "dairy cities" zoned for agriculture with low taxes and few city services.[81] The dairy cities were one response to the San Joaquin Valley challenge to their market. In this way, peri-urban producers were able to reduce their costs of production and maintain their access to milk markets. These Dairy Cities also enabled milk processors to maintain cheap supplies in peri-urban areas, because of low assembly costs. Yet, they also held the potential of multiple sourcing over the heads of these farmers: "If producers in one area should bargain collectively for better contracts, the processing firms can hold the threat of transferring contracts to the competing area."[82] As a result, "[t]he two producer groups were separated and antagonistic toward one another, and the major milk distributing companies reaped the benefits."[83] The California struggles over access to the Los Angeles market in the 1950s repeat the history of sectionalist dairy politics already described in New York. However, unlike New York farms, peri-urban Los Angeles farmers were able to expand production—through an increasingly intensive dry-lot dairy system—supplying a year-round and relatively inexpensive milk for the urban populace.

California's diverse agricultural economy also influenced its dairy policy. Unlike rural areas in Wisconsin and to some extent New York, rural regions in California were not solely dependent on the dairy industry for their income. Regions that went out of dairying could often produce other crops. Therefore, state and city officials could either wholeheartedly support, or simply not interfere with, the political machinations that made peri-urban, large-scale dairy farming economical without fear of politicians running to the rescue of their middle-ring dairy farm constituency. Rural politicians in California, unlike those in Wisconsin and parts of New York, were not faced with the prospect of representing districts that were increasingly populated only by bears and trees (or lizards and sagebrush, none of which are eligible to vote) if dairying left the area.

In California, as in Wisconsin, the method of pricing surplus milk has historically strengthened the protective boundary between market and manufactured products by strongly restricting entry into the fluid milk market. California has a state market order system, which gives the state greater flexibility in setting prices and managing markets. As a result, California was able for many decades to manage the boundaries between various dairy systems within that state with a unique regionally based "cost plus" pricing system similar to utility regulation.[84] For many decades, fluid milk "dairymen"[85] near and in the city of Los Angeles were protected from the cheaper milk further away by the fact that the Los Angeles regional market orders set both higher prices for dairymen and higher prices at the retail level.[86] Until the 1970s, retail price setting meant that the industry was able to pass these higher prices onto consumers and therefore dairy businesses were less tempted to go out of the area in search of cheaper milk.[87] Most important, since the 1930s, a quota system has restricted access to higher-priced fluid milk markets. Historically, most of this quota has been "owned" by fluid milk dairymen on "ranches" near cities.[88] Originally dairy processing plants owned the allotments, but they were transferred to farmers in the late 1960s. Farmers continue to buy and sell these allotments today.[89]

Unlike officials in Wisconsin and New York, California agricultural officials did not follow any overt policy concerning cooperatives.[90] Cooperative organization and management, like most agricultural policy, was left to farmers and their organizations. Agricultural institutions played more of a service role to farmers as opposed to the more proactive policy roles they played in Wisconsin and New York.[91] However, as in Wisconsin, cooperatives, before current mergers, were generally local and divided between fluid farmer bargaining associations near cities and manufacturing cooperatives in areas far from the urban market. Inner ring California producers, on the other hand, have belonged to what are referred to as "captured co-ops," primarily bargaining associations tied to a particular company.[92]

Like New York and Wisconsin, California instituted a program to study land utilization, and investigated the problems of scattered settlement of state population. A major study focused on the foothills of the Sierra Nevada. In this area, incomes were low and many people were on relief, due in part to the decline in the mining industry. According to one project researcher, Howard Conklin, the goal of the study was to explore the issue of where the lower boundary of the forest should start, and human settlement end. Land planners referred to this problem as determining "the margin between livestock ranching and forest."[93] Emotions over the answer to this question ran high, making it "a shooting matter" in the local communities studied.[94] The question of determining the proper boundary line between forest and extensive agricultural uses such as ranching involved determining which areas were "marginal" and whether these marginal land uses should be discouraged or eliminated.

However, unlike the New York state land utilization study, the Sierra Nevada land utilization report did not recommend the removal of settlement from these areas, for a number of reasons. First, removal of settlement would not have a significant effect on road maintenance "While elimination of scattered settlement probably would result in some reduction in expenditures for roads, economies might not be considerable since the roads serving these populations usually would not be abandoned because of uses for recreation, fire protection, or other purposes."[95] In addition, the Sierra Nevada report, unlike the New York report, acknowledges the contribution of subsistence farming and other noncash receipts to total income. The study found that noncash activities contributed nearly half of all income in the areas studied. Taking into account the contribution of subsistence farming to total income, the report argues that the maintenance of settlement in the area may actually lower total state expenditures for relief: "It is possible, however, particularly when widespread unemployment exists, that maintenance of populations on subsistence ranches in relatively sparsely settled areas may be less expensive than and preferable to concentration in better but more densely populated areas where living costs are higher."[96] Instead of recommending removal of population from the upper foothills, the report advocates "improving the plane of living of the population through a more economical and complete utilization of resources" available, thereby "strengthening the sources of nonfarm as well as of farm income." This included not only "the maintenance of forest employment" but also "the solution of range improvement problems."[97]

Why was it that, given the similarity of problems in the marginal land areas of New York, Wisconsin, and California, the policies recommended to deal with these areas were so different? The main reason was the need for New York planners to remove from the market milk produced "at a loss." The increasing competition between the two dairy networks, which New York dairy officials could not control through the establishment of impermeable market boundaries, required instead the elimination of more extensive summer dairy farms.

THE INDUSTRIAL PRODUCTIVIST VISION TODAY: rBGH

The activities of Cornell agricultural economists and animal scientists today continue in pursuit of the industrial vision. Ironically, New York state's dairy farms are still not highly industrialized, particularly as compared to dairy farms in the West and South. Yet Cornell economists continually portray these western farms as the ideal farm of the future that will outcompete other forms of dairying.[98]

In particular, Cornell has played a strong booster role in support of a controversial biotechnology that raises milk yield: recombinant bovine growth hormone, or rBGH, a bioengineered substance that increases a cow's milk production even more than what has been accomplished by decades of breeding.[99] Early assessments predicted a rise in milk production of 10 to 25 percent in cows injected with this genetically engineered substance. The close collaboration between Cornell animal scientists and agricultural economists with Monsanto during the assessment of bovine growth hormone indicates that the pro-industrial politics of this land grant institution remain strong. Dale Bauman and other dairy experts at Cornell have been at the forefront of rBGH testing on cows, and have vociferously defended the safety of this product against a large consumer backlash.[100] Interestingly, a Dale Bauman press photo during the 1980s updated the traditional expert image, with the white-coated scientist injecting rather than inspecting the cow.

Agricultural economists at Cornell were among the first to predict wide adoption of rBGH and massive loss of dairy farms due to the increase in production that would result from this new technology. In their assessment of the economic impact of bovine growth hormone on the dairy industry, these economists predicted that the new technology would force many small farmers out of business, because the technology requires intensive herd management and "[l]arger herds are indicative of better managers."[101] As a result, the appropriate policy response recommended was "to encourage the orderly exodus of resources, including farmers, from dairying."[102]

In fact, due to consumer concern, cow disease, and reproductive problems, many farmers who once used rBGH have "de-adopted" the technology.[103] Less than a third of all farmers use this milk-inducing substance today. In addition, the substance has not lived up to expectations: earlier predictions of a 25 percent rise in production have not been fulfilled.[104]

Yet not all state legislatures and their client land grant universities have pursued an industrial vision for dairying. Ironically, it is California, the state built on industrial agriculture in so many other sectors, that has most actively supported diversity in the dairy system through its state milk market order. While California has some of the most industrialized dairy farms in the country, it also has a history of milk pricing that distinguished dozens of regional marketing areas and based prices on cost of production within these areas. As a result, until the 1990s California had a pricing system that supported dairy strategies based on smaller farms, pasture-based feeding, or mixed farming in certain regions of the state.[105] In the 1990s the merging of the California regional orders into two big orders, and the move from cost of production to formula pricing, dampened this regionally specific regulatory system and made it more like the federal market order system. Even so, California still has a large number of dairy producers who rely heavily on pasture.[106]

Rather than collaborate with biotech companies during the rBGH assessment period, as generally occurred at Cornell, agricultural experts in California have been remarkably muted in their rhetoric. One interesting product of this agricultural system was a study comparing the adoption of rBGH to the adoption of an intensive pasture system.[107] Although the study has been highly controversial, it reflects a difference in approach from the pro-industrial dairy forces at Cornell.

In Wisconsin, the hiring of Frederick Buttel in the 1990s to direct the state's agricultural technology center is an indication of a different approach to the biotechnology issue in the state. Buttel has been the major land grant critic of biotechnology and of the university-industry collaborations that have occurred in this high-tech sector in recent years. His assessments of rBGH have been highly critical of Cornell's optimism about the technology.[108]

CONCLUSION

This explanation of agrarian politics and dairy policies in three states shows that the dairy economy is not the product of inevitable, autonomous forces. Industrialization is as much due to social vision and political contest as it is to impersonal forces. The "milkshed" as a spatial phenomenon was itself a product of the cultural move toward fresh milk drinking and the growing power of urban constituencies. Later, consumer politics was essential to the creation of a productivist dairy policy. In each case, dairy production has been shaped by a history of political bargaining. State, business, consumer, and farm interests aligned with each other in different ways in the three dairy states, based in part on differences in their spatial and economic positions.

The industrial bargain struck in New York state forged the proactive, state-encouraged form of productivist policy that was the precursor to agricultural development policies later carried out in Third World countries. New York's form of productivist state intervention eventually became the "cheap food" agricultural development policies promulgated by the World Bank and other development institutions over the last half century.[109] Politicians choose cheap food policies in response to the growth of cities, which become pools of labor and votes. Urban politicians in growing Third World cities have needed to keep their urban constituencies fed cheaply, so as to attract capital investment seeking low-wage labor. From this policy viewpoint, the purpose of rural areas is to provide the labor and the cheap food to feed that labor.[110] The country, at best, becomes merely a source of energy for the urban machine. At worst, it is forgotten altogether. Development experts attempt to transform the countryside to this vision of intensive production, the Green Revolution being the most successful of these efforts. Social change occurs, and is directed by, the city.

This should tell those of us interested in alternatives to the current food system that we need to look more closely at the political cultures involved in the creation of economic environments that allow for production alternatives. These social factors should include the overwhelming power of consumers and the bargains struck between consumers, politicians, industrialists, and farmers.

Yet most policies to create a more diversified agriculture simply concentrate on discovering new, and preserving old, ways to farm. Trying to create a new form of food production by simply trying to change farmers' practices will not make sustainable, diversified, or alternative agriculture viable. As this retelling of the story of milk has shown, to reenvision new ways of producing our food will require a renegotiation of the social and political relationships between urban, suburban, and rural areas. Without attention to this larger institutional, political, and relational context, the ability to change our food system is lost. It is time to rethink the industrial bargain and to reorganize the production of our food on new terms.

10

The End of Perfection

AMERICANS LIVE IN a world filled with milk. Producing 153 billion pounds a year, the United States is "the largest milk-producing country on the planet."[1] Many of our most popular entertainment, sports, and political figures have appeared wearing the telling milk mustache. Milk promotion money, deducted from farmers' paychecks and flowing to the dairy councils, saturates us with advertisements. Everywhere we hear the ominous question: Got Milk?

Yet milk fills our world in another way as well. Once again, this food is at the forefront of an intensive debate about American forms of eating. In part, this is due to a quirk of genetics: the hormone that increases lactation is particularly easy to biotechnologically replicate. In most cases, a large combination of genes controls bodily activities. However, only one gene in a cell regulates lactation. This same gene regulates growth and is referred to as "growth hormone." Growth hormone was therefore the first commercial biotechnology product.

Yet, whom to sell it to? Selling parents on the idea of making their children taller has had only limited potential, so far. However, the biotech industry quickly figured out that the most profitable potential product was bovine growth hormone. Injected with a recombinant form of the growth hormone cows normally make when lactating, cows increase milk production. After a contentious FDA approval process, recombinant bovine growth hormone (rBGH) came onto the U.S. market.

This first commercial biotechnology product was also the first political target in the contentious debate over genetically engineered foods. One observer has referred to rBGH as "the prototypical politicized agricultural technology of the late twentieth century."[2] The struggle over rBGH has more recently expanded to soybeans, corn, and other agricultural commodities. Current assessments estimate that more than two-thirds of American food contains some form of genetically engineered ingredient.[3] A growing social movement to ban genetically engineered foods from the market and a concomitant consumer boycott of these foods have gained a great deal of public attention. Both the European Union and Canada have voted against allowing genetically engineered crops into their marketplaces. The debate over rBGH has been at the forefront of all these political discussions. As a result, rBGH has been of central importance in the formulation of a story in opposition to genetically engineered food.

The anti-rBGH movement, however, quickly segued into a larger conversation about milk itself. For the first time in the nation's history, the perfection of milk has come under public scrutiny. For example, a burgeoning number of Web sites with titles like "notmilk" and "milksucks" have appeared on the Internet. Previously, criticism of the milk system focused on political corruption and economic monopoly. These earlier critiques focused on big business and bad government preventing the public from getting pure milk at a fair price.[4] Only with the advent of rBGH has this food itself come under question. The long-assumed perfection and indispensability of milk are no longer part of American iconography. For many, "Is our milk safe?" is no longer the question, but "Should we drink milk at all?" One growing food-based social movement, veganism, argues against the consumption of any dairy products.

Joining this new questioning of milk are those groups of people who tend to be genetically lactose-intolerant. The Congressional Black Caucus, for example, recently challenged the prominent role milk plays in the USDA's food pyramid, a graphical view of the agency's nutrition recommendations.[5] The caucus wrote a letter to President Clinton asking him to consider changing these dietary guidelines to put less emphasis on milk.[6] Yet the new dietary guidelines released in June 2000 did not revise the food pyramid. Instead, the text now advises those who "choose" not to eat dairy products to make sure to get calcium from other sources.[7] While the text notes that some people may make this choice because they are lactose-intolerant, there is no attempt to educate people who belong to ethnic groups that tend to be lactose-intolerant that they might consider choosing alternative ways of getting enough calcium in their diet. In other words, the food pyramid is still for everybody, that is, *every body*.

An increasing number of scientists and scientific organizations have questioned the emphasis on milk in American nutrition policy.

These scientists tend to be the less mainstream Physicians Committee for Responsible Medicine, rather than a much larger number of scientists and scientific organizations that support milk drinking, such as the American Council on Science and Health and the American Dietetic Association. Each of these groups accuses the other of being tied to special interest groups. Whether or not this is so, the rBGH controversy has made criticism of milk more audible in the media. A prominent anti-rBGH Web page by the Organic Consumers Association contains links to anti-milk pages as well.[8] In other words, Monsanto and its rBGH product may have had more impact on decreasing milk consumption than the millions of dollars invested in campaigns to increase milk consumption.

The anti-milk narrative turns the perfect story on its head. Everything that is true in the perfect story about milk becomes untrue in the anti-milk story. In contrast to the perfect story, it is a "downfall story," a reversed narrative that paints history as the story of decline. Everything celebrated in the progress history becomes another benchmark in the history of human downfall.

In the downfall story, previous forms of food production that are simpler, more local, less industrial, and more craft-based are often romanticized as lost, ideal forms of producing food. For example, Jim Hightower's article "Food Monopoly" in the Naderite *Big Business Reader* begins,

> Americans have come a long way since the first Thanksgiving meal was brought to some New England colonists by local Indians, in a spirit of sharing, 367 years ago. Maybe we've come too far. At least those Pilgrims knew where their meal came from—it was the bounty of nature, delivered by Indian people, and presumed to be the blessing of God.
>
> No longer is that the case. These days, nature has less and less to do with our meals, which are often not even put on the table by farmers, much less Indians; and if modern food is God-ordained, as Earl Butz once suggested, then the religious fundamentalists are right—we've ticked off the Lord something awful. His revenge is a Brave New Thanksgiving that, unbeknownst to most Americans, is the product of monopolized markets, conglomerate bookkeeping, genetic engineering, integrated factory systems, centralized procurement, national advertising, chemical artifice, standardized taste, and The Bottom Line.

> It's not especially good, or good for you, and it's very expensive, but you can be thankful for one thing: there's plenty of it.[9]

From the downfall perspective, the modern food system and state regulatory systems "captured" by modern big business interests have degraded the quality of our diet, making it less God-given. The decrying of big business "trusts" or "combinations" of the earliest city milk dealers, the later large dairy corporations, and the current genetic engineering companies, have all followed the downfall story. The downfall story points to modern society, technical progress, and public institutions as the source of social problems, not the solution. The current downfall narrative is centered on the chemical/biotechnology company Monsanto and its manipulation of government and the media to force onto the market milk created with rBGH.[10]

Of course, like the perfect story, downfall stories contain a great deal of truth. The downfall narrative has provided a crucial critique of modern triumphalist progress narratives, giving the public a clue that not all is well with the food they eat. Downfall narratives focus on those who profit financially from the selling of milk and how these interests have taken advantage of the public.

While the downfall story helps us gain an understanding of the food we eat, and of our modern world in general, this story approach has a number of weaknesses. By making the goal of critical storytelling simply the turning of triumphalist narrative on its head—everything that is good becomes bad—the downfall narrative does itself a disfavor. Downfall narratives no less than progress narratives are stories of perfection. Through a politics of reversal, what all progress narratives say is purely "white" becomes purely "black." The downfall narrative therefore chains its story to the story of triumph, rendering it incapable of seeking and articulating alternative routes from the past and to the future. This limits the sort of questions that can be asked about history, specifically in this case about milk. As a result, there has been little examination of the workings of the deliberations themselves: the politics, the rhetoric, the creation of meanings, and how each political actor in this public discussion both affected and was affected by these doings.

Certainly, the downfall story and the politics of reversal have been powerful political weapons for those who actively question the perfect story. As Antonio Gramsci noted, to act politically, people need to have a web of meanings to work within. Political action without meaning is impossible. Yet a web of meanings is not possible without a story. We all need stories about who we are and what our society is in order to try to change ourselves and change society. Since people cannot act politically outside a set of meanings, it is necessary for those who want to act outside the perfect story to have another story to tell.

The downfall story has provided that set of meanings for many people. Yet, as a perfection story in reverse, it is also problematic. The perfection story tends to present one set of people as perfect and others as inferior. Perfect stories tend to denigrate diversity and dispersion and encourage conformity and centrality. The Cold War is an excellent example of a situation in which two perfect stories, each claiming to be the exact opposite of the other, resulted in enormous amounts of centralized authority and social conformity, on both sides. The duality of the progress and downfall narratives tends to get us caught in political quagmires that the media love to exploit: abortion, school prayer, crime are all presented as polarized issues, each with its televised talking head representing one side of the question on the evening news. As soon as a political position has a perfect story, open discourse for positive change becomes nearly impossible. There is no opportunity to present a different story besides the negation of the perfect story.

The politics of downfall therefore has all the problems of a perfect politics. This book has mostly critiqued the perfect story and its dependence on ideas of universality, completeness, hagiography, and public enlightenment as ways to improve society. The downfall story depends just as much as the perfect story on these ideas. Therefore, the downfall story suffers from all the problems of the perfect story as related throughout this book. The downfall story will not get us to a more social story, one where diversity and contingency, localness, human choice, and imperfection play a role.

As we have seen throughout this book, each form of politics has its story. A story that paints the practices of one group of people as perfect creates a politics that automatically leaves some people out of the decision-making process and puts other people in authority. The opposite story, of downfall, questions the authority of those in charge of the perfect story, yet it cannot formulate an alternative vision that is unencumbered by that story. As a result, in both cases, the political imagination gets cut off and social movements end up being sideshows to the major story of progress, today painted as the promise of global markets. Both progress and downfall milk stories fail to place milk in its actual social context. Either perspective treats the advent of the modern food system as an inevitable march of progress or perdition, helped along by reformers and scientists.

Are we doomed to simply repeat the downfall and progress stories and their politics? This final chapter will look at the new politics of milk to discover whether or not we can find a more social, less-than-perfect form of politics. Accordingly, this chapter will examine two anti-rBGH movements and their attempts to rethink milk as an American food habit. The anti-milk movement completely rejects milk as a food. It presents milk as a degenerate product and the rise of milk drinking as a downfall narrative. The anti-milk movement, including veganism, creates a powerful community of practice, yet its downfall narrative makes it problematic as a form of politics.

The anti-rBGH movement is also critical of the current food production system but does not reject milk as a food. Instead, this movement has supported the rise of rBGH-free and organic milk. Yet re-creating a food is more complicated—and more social—than rejecting it. As we will see, the rise of organic milk consumption represents a more diverse, less perfect form of politics. Social resistance to genetic engineering happens not just in the realm of social movements but also in the seemingly more individual, private realm of consumption.

The rise of organic and rBGH-free food consumption is often denigrated as simply a neoliberal form of "false" politics, based on the agglomeration of individual rational decision making and catering primarily to the status-conscious upper-class.[11] In contrast, this chapter will present this new form of consumption as explicitly political: a less organized but still significant form of politics in which practice and identity are being bargained out on a day-to-day basis.

THE ANTI-MILK MOVEMENT

One example of the anti-milk point of view on dairy is Robert Cohen's 1997 book *Milk: The Deadly Poison.*[12] This rambling tome is a compilation of every critical statement ever made about milk. He garners data showing that milk causes breast, colon, and other cancers, heart disease, diabetes, early maturation, and osteoporosis, the brittle bone disease that milk drinking is supposed to prevent. Cohen even carries out an intricate but strange examination of the athlete Florence Griffith Joyner's autopsy, in an attempt to show that she in fact died as a reaction to milk consumption. On his Web site he refers to the somatic cell bacteria present in all milk as "pus," even reversing the milk mustache advertisements with counter-ads of milk "pus-staches."[13]

Cohen's book is a classic downfall narrative, taken to the extreme. Everything commonly accepted as good about milk is, according to this narrative, actually the reverse. The prescription is to drink no milk whatsoever.

Because downfall narratives are politically problematic, they tend to gain a small but loyal group of followers, people with lives that are open to conversion. Single people from privileged backgrounds tend to populate these movements, not because these movements are necessarily elitist, but because most other people are constrained by the social networks they find themselves in.[14] The complete rupture necessary to change from the common American diet to a vegan, no-milk diet requires renegotiating the rules of life with other people, a difficult political process the downfall narrative makes even more difficult by predetermining right living as the perfect opposite of perfection.

Yet veganism can also provide its adherents with the tools to overcome the constraints of particular life habits, providing a strong and unified alternative "community of practice" that enables some people to make radical changes in lifestyle. Earlier, we used the term "community of practice" to define the type of social context that enables most women to breast-feed and the ability of dairy farmers in some regions to practice a more diversified form of agriculture. Yet all forms of eating require a community of practice, a particular way of life that embeds a person in a network of people who support that practice, so that the "performance" of that practice leads to satisfaction and self-esteem. If your everyday practices expose you to significant social censure and denigration, it is difficult to continue those practices. This is particularly true if you lose the economic network to make a living. Each set of everyday social practices, to survive, needs a community that values those practices. Vegans find that community in other vegans. In fact, at my home campus, University of California at Santa Cruz, veganism is almost the norm among counterculture white students. It belongs with a set of white counterculture practices such as drumming circles, rasta-style hair, bicycle-commuting, and hanging out in front of the Wednesday Farmers' Market.

Even Weight Watchers depends on a community of practice to help members change their everyday habits. Membership in this group links you to a British duchess, and turns the loss of a few pounds into a community celebration. It is this creation of a community of practice that makes Weight Watchers effective.

Of course, the perfection narrative has created the most powerful community of practice of all. It is what Gramsci would call a "hegemonic" community, a set of habits supported by nearly every American institution, making the creation of alternatives extremely difficult to imagine, much less implement.[15] Our current American food habits make us rely on milk to provide the vitamins and minerals that are absent in the other foods we eat. As Americans, seldom have we ever heard advertisements celebrating the benefits of kale. There is no American Kale Council supported by a percentage of every bunch of kale sold in stores or farmers' markets. *And there never will be, despite the fact that kale is just as good for you as milk.* The problem with milk is not that it is bad for you, but that it has a whole institutional apparatus that has made it *the* celebrated food, when many other foods and many ways of eating are just as deserving.

The perfect and downfall stories also create a perfect politics: either you are a loyal milk drinker or you don't drink it at all. On the one hand, you are healthy from a governmental point of view, on the other hand, you are "healthy" from the viewpoint of a certain type of political correctness. But the vast majority of people find both government officials and politically correct people annoying. Both make us feel policed and apologetic in our everyday activities. I always cringe when I think of my vegan students catching me waiting in the drive-in line at Burger King. Vegans are right when they say that animal factory farming is cruel. It doesn't take too many pictures to convince the average person of that. But try to negotiate that position with children embedded in a Happy Meal community of practice ("What toy did you get?").

For most people, everyday practice involves a continual bargaining, a continual compromise and "satisficing" between what we would like to do and what we have to do. Who we are is intricately caught up in the web of what we do and what we do is caught up in the web of our relationships with others. Therefore, our everyday practice is part of the everyday negotiation of our identities and the identities of the communities we live in.

FAT, FAMOUS, SEXY, AND ADDICTIVE: DAIRY ADVERTISING

Everyone knows that the double bacon cheeseburger is not a healthy meal, and it isn't advertised as such. Fast food chain restaurants advertise a fat-inducing, artery-clogging meal as a risqué adventure into pleasure. These foods are most commonly eaten by what the fast food restaurant industry refers to as "heavy users": single white men between twenty and thirty-five.[16] Members of this group revel in the pleasures of industrial eating, and die from heart disease at a much higher rate than those in other groups. Presumably, members of this market niche see themselves as adventurous risk takers. Burger chains therefore market these unhealthy meals in ways that enable these men to reinforce their identities. These ads are a form of identity politics.

Dairy ads have jumped into identity politics as well. The dairy industry has discovered the identity struggles that are part of everyday practice. Its marketing strategy associates milk with positive identities. The milk mustache ads are an excellent example of this: celebrities are caught in the act of drinking milk, as evidenced by the milk residue on their upper lip. This hegemonic community of practice communicates a linking of esteemed identities with the practice the ad seeks to encourage. In this way, a person gains satisfaction and self-esteem by participating in this same practice. This dairy marketing strategy is not new. Silent movies in the 1920s portrayed baseball players and beauty queens drinking glasses of milk with gusto.[17]

Yet there is one significant difference between these ads and the 1920s versions. The old milk ads portrayed one form of physical perfection: white beauty and athleticism. The new milk mustache ads instead portray a wide diversity of people in terms of race and social background, including the comedienne Whoopi Goldberg and the Clinton cabinet member Donna Shalala, both of whom have more than a 50 percent chance of being genetically lactose-intolerant. Yet in the eerie similarity of each picture, the voice of authority ("Got milk?") implies that milk drinking is a universal, everyday practice, no matter what your race, culture, or social background might be. The image is a classic example of the politics of perfection.

A more contemporary, yet more ominous, National Dairy Council ad involves a boy who leaves home because he is addicted to milk. His mother reads letters from him, traveling the country, binging on this food. A less outlandish ad involves the problems of being without milk when you find yourself in the presence of a cookie. These ads play on the tacit, less rational urges that are part of a community of practice: our need for milk may not make sense, but we love it anyway, because it defines us. Because the milk industry is constantly trying to beat out that urge to grab another American food—soda—dairy ad executives try to make milk an exciting, crave-inducing beverage.

Vegans are not the only Americans currently questioning this perfection story. The recent but spectacular rise of organic milk as an industry represents one product of this questioning. Yet, the politics around the rise of organic milk are ambiguous and contradictory. The following analysis of the rise of organic milk shows that, in fact, this consumer decision making is political, but it is a less than perfect politics. To understand this politics requires a rethinking of the role of the consumer in the food system. In this case, a reformulation of consumer organic milk consumption as a "Not-in-My-Body" or "NIMB" politics of refusal brings the political aspect of this consumer activity to light. This form of politics is similar to the refusal of neighborhood residents to accept toxic facilities in their neighborhoods, generally referred to as "Not-in-My-Backyard" or "NIMBY" movements. A comparison of NIMBY and NIMB will show that both forms are politically "reflexive" even though NIMBY involves a more conscious and organized social movement. To reveal the nature of NIMB politics requires attention to the ways people talk about milk today. One place where producers talk about milk is on their milk cartons. A comparative analysis of organic and nonorganic milk carton discourse will reveal the nature of milk's NIMB politics.

rBGH AND THE RISE OF ORGANIC MILK

"It's exploded," says Joe Smillie of Quality Assurance International (QAI), a San Diego-based company that certifies organic products. "We're seeing more dairy products come through for certification now than anything else. It's obviously today's hot button."[18]

This quote and others from the trade magazine *Natural Foods Merchandiser* describe the phenomenal rise of organic dairy products. The article that contained this quote, titled "Organic Dairy a Cash Cow," notes in particular the rise in sales of organic fluid milk.

The statistics on organic milk's rapid rise are impressive. Financial analysts estimate sales growth to be 50 to 80 percent annually, with total sales of approximately $60 million in 1998.[19] While organic milk is only a small fraction of the $75 billion American dairy market, market analysts predict that the strong rate in sales growth will continue, possibly reaching 2 percent of the market by 2002.[20] In contrast, sales of organic products as a whole are growing approximately 20 percent a year.

Why this explosive growth from virtually nothing less than a decade ago? Trade journals and the mainstream media all mention one reason: rBGH, a genetically engineered hormone injected into cows to increase their milk production. Produced by Monsanto under the commercial name Posilac, this substance has been controversial since the Food and Drug Administration approved it in 1993. Nearly every popular and trade press article on the rise of organic milk mentions rBGH as the main reason consumers changed their consumption practices. For example, a January 1999 *New York Times* article quotes one mother saying, "Milk is such an important part of a child's diet. . . . I didn't want my child to be a guinea pig."[21] A *Detroit News* article describes a natural foods store owner running out of organic milk on a regular basis: "Customers tell her they're buying the milk because of concerns about hormones and antibiotics in standard milk."[22]

Yet organic milk has another unique aspect: unlike the demand for most organic food, consumer demand for organic milk arose without the significant social and political organizing—the food co-ops, the consumer-farmer coalitions—that created the organic food system over the last few decades. Many of the more mainstream organic companies—providing produce, grains, and processed products—began as small-scale firms dependent on a politically or nutritionally aware consumer who purchased food in alternative marketing channels such as co-ops or food clubs.[23] Since then, many of these alternative marketing organizations have been replaced by private stores and many of the products have moved onto the shelves of more conventional food retailers, such as supermarkets. In contrast, organic milk did not go through this transitional "hippie food" period, in which alternative food retail organizations and politically conscious consumers provided the incubator environment for industry growth. This unique aspect of its market has attracted trade industry attention: "People who don't buy any other organic products are purchasing organic milk," states Katherine DiMatteo, executive director of the Organic Trade Association.[24] Organic milk is a "crossover food," infiltrating markets other organics don't reach.

MARKET NICHE OR POLITICAL ACTION?

The fact that many of its consumers are "mainstream"—that is, they do not see their consumption of organic milk as part of any wider social movement—makes it easy to dismiss the rise of organic milk as simply another upper-middle-class "niche market." In fact, recent work on the organic food system challenges the industry's self-image as a philosophically alternative economy.[25] From the perspectives of these studies, organic food production is simply another form of "postindustrial" capitalism, a new economic bargain between (certain) consumers and industry in which large-scale "business-as-usual" capitalist enterprises become more flexible in meeting a more diverse set of consumption demands. The fracturing of consumption has created various market niches, which include a demand for a form of food defined as "organic." To provide this "postindustrial" organic food, producers find ways to meet certification standards while maintaining large-scale, intensive, less environmentally sustainable forms of agricultural production. The postindustrial organic food critique challenges the philosophy of the natural foods movement, which holds that consumers can change for the better the way food is made by demanding organic products.[26]

The difference between these two perspectives centers on the role of the consumer in the structure of food provision. Those who see organic foods as a social movement make the consumer an actor in the politics of food.[27] In contrast, the postindustrial organic perspective takes a Marxian point of view, locating politics in the relations of production that manipulates consumer demand for the purpose of greater profits. In the case of organic food, the postindustrial firm will respond to consumer desires for particular food characteristics like "quality," "made with care," and "good for the environment" by mimicking the types of firms commonly associated with these characteristics, namely, small, artisanal firms and family farmers. The consumer, in this framework, is either a victim of false consciousness or is part of an elite class of people who eat particular organic products as a way of displaying their cultural capital.[28] For example, one organic food often denigrated as a status food is "salad mix," termed by Buck et al. as "one of the top yuppie commodities of the nineties."[29]

This Marxian analytical reading of organic food consumption is part of a larger framework commonly known as food system or "commodity chain" studies. The food system/commodity chain approach studies the institutional structure of food provision from producers, distributors, processors, transnational corporations, government agencies, and international trade organizations to the consumer. This perspective on agriculture and food, however, tends to approach the "consumption side" of food systems from a Marxian structuralist perspective: how we eat is a product of larger, macroeconomic forces. Goodman and Redclift's characterization of food consumption after World War II, "Food into Freezers, Women into Factories," exemplifies this view, making changes in the food system part of an overall structural move bringing women into the labor force.[30] Mintz's study of the rise of British sugar consumption, despite its subtle weavings of class and culture, also attributes this change in the working-class diet to changes in class relationships due to the rise of the industrial workforce.[31] In this case, food is a product of forces, and eating reflects these forces. Recent studies of the post-Fordist food system have portrayed new consumption practices as the product of a bifurcation of classes into (1) professionals concerned with health and the status of eating artisanal foods and (2) "everybody else."[32]

From this point of view, the consumer is not a focus of theoretical attention because she or he does not "act." The definition of action, in this case, is collective, politically conscious action. For example, Marsden and Wrigley, in an article on the retail food industry's increasing control over the British food market, conclude by stating that the extent of consumer action in the future "will depend upon the development of the social and political consciousness of the consumers themselves. This in turn depends upon an ability to overcome the types of commodified individualism and positionality much of the contemporary system attempts to promote."[33] Consumers have an "undeveloped" consciousness, which will continue to be undeveloped—that is, unpolitical—until they challenge the commodity system. Likewise, the *Nation* disparages any concept of consumer as actor, calling all such activity "tofu politics," a degraded form of action in which individuals believe they can make a difference not through collective action but through shopping.[34] This denies any politics to the act of consumption at all.

Despite its lack of attention to the consumer, the food system analysis of organic food describes and explains a real trend in the industry. For example, Buck et al. provide an excellent analysis of how post-Fordist forms of organic production are furthered by an organic certification system that emphasizes specific "standards"—in terms of the toxic inputs not used—and de-emphasizes the proactive "process" criteria of agroeconomic farming, such as care of the soil.[35] This commodity chain approach furthers our understanding of how a set of standards, originally formulated from a philosophy of small-scale, local production, can be captured by large-scale multinationals.

BIG BUSINESS ORGANIC?

One major difficulty in finding the politics in the rise of organic milk is the fact that a major part of the organic milk industry arose, fully grown, from the heads of corporate executives and Wall Street investors. The organic milk industry is *more* concentrated than its conventional counterpart. If the claims of the top five companies in this sector are true (and they may have hyped their claims a bit to convince Wall Street analysts), the five companies listed in table 10.1 serve 95 percent of the organic milk market. In fact, this table understates industry concentration, since the data were collected before Horizon purchased both the Organic Cow and Juniper Valley firms, leaving only three firms—Horizon Organic, Alta Dena, and Organic Valley—serving 95 percent of the organic milk market.[36]

Horizon buys some of its milk from local organic farmers. However, 75 percent of its milk comes from two mega-dairy farms in Idaho and Maryland. The company bought the Maryland and Idaho herds, 5,400 cows, of Aurora Dairy, a multistate dairying corporation that, until the sale, was the largest dairy producer in the country.[37] Horizon, in other words, is a vertically integrated dairy company, providing organic milk mostly by transporting it from its own two centralized herds numbering in the thousands of cows. In contrast, most conventional dairy companies—even the largest national firms, Suiza and Dean Foods—do not tend to own farms. They buy their milk from producers, mostly within the home market region.

Table 10.1

Claimed Market Shares of Five Largest Organic Milk Companies

Company	Share
Horizon Organic	65
Alta Dena	10
The Organic Cow	9
Organic Valley	6
Juniper Valley	5

Source: "Organic Opportunities."

Horizon Organic Dairy's fresh milk business is also more national and multinational than most conventional counterparts. It is a *global* organic milk company, having just entered the Japanese market.[38] It is partially owned by Suiza Foods, the second largest dairy processor in the United States, which currently owns a number of regional milk firms. While Suiza and Dean Foods are certainly larger companies, Horizon is currently the only U.S. milk *brand* (organic or conventional) that is national. Anyone familiar with the history of commercial food in the United States knows the importance of "branded" foods: nationally advertised products, with distinctive containers and centralized production systems. While branded foods have been with us for a long time, economic analysts note that, in this postindustrial food economy, branded foods have become the only types of food products large publicly traded corporations want to own. It is Horizon's ability to create a national organic milk brand that makes it unique.

Nationally branded milk corporations are not new in the United States. Borden's and Sealtest were once national brands. Yet both of these companies withdrew from the fresh milk business in the 1970s. Alden Manchester, the USDA's dairy marketing guru, explained the national brand sell-off of conventional milk businesses by multinational firms: "[m]anufacturers of consumer goods derive much of their market power from product differentiation through brand preference." Large, publicly traded companies have mostly left the conventional market milk industry to cooperatives, because "Wall Street now favors high-margin branded products. Commodity lines, which include fresh meat, fluid milk, natural cheese, canned fruit and vegetables, and raw sugar, were sold off rapidly in the 1970's by companies wishing to specialize in branded foods."[39] Milk and most other U.S. dairy products are "bulk commodities" because they are generally produced according to national standards. As a result, most milk and milk products taste the same from one place to the next, making product differentiation difficult. Because "[r]eal differences in flavor, texture, or quality are extremely helpful in creating brand preferences, . . . [t]he creation of strong brand preference has never been easy for many dairy products."[40] The brand gives large corporations access to a profitable market niche in the fractured consumption society. Yet today, two large multinational food companies, Suiza and Dean Foods, are purchasing local milk brands and leaving on the local label. The companies in this case have purchased companies with strong local consumer preferences and hope to retain these preferences by keeping local brand names.

Yet there is one small light in Manchester's assessment of the future of dairy marketing: organic and non-rBGH milk. This product represents one of the few opportunities dairy producers have "to market differentiated identity-preserved products."[41] In other words, the anti-rBGH movement—and, indirectly, Monsanto itself—has provided organic milk with its identity, exactly the kind of high-margin product identity necessary for the product to become a profitable Wall Street investment. Because of its ability to create a national brand differentiated by its organic certification, Horizon attracts investment capital.

One of the largest dairy processors in the United States is Dean Foods, which owns the second largest organic milk company, Alta-Dena. Alta-Dena, a longtime player in the Southern California milk market, has been the subject of various sales and mergers. Dean Foods acquired Alta-Dena from Bongrain North America (a large French multinational). Therefore, both Horizon and Alta-Dena are owned by publicly traded, multinational firms. Ironically, therefore, consumers buying organic milk may be passing over a smaller, cooperative, regional product to reach for the multinational, publicly traded organic product.

Organic Valley (listed fourth in the table) is the brand name for the Coulee Region Organic Produce Pool (CROPP), a 160 farmer–owned cooperative in Wisconsin that sells its milk in thirty-two states, mostly in the Midwest market.[42] It has also recently expanded its membership to the West and Northeast. It supplies milk to both Horizon and Alta-Dena. Compared to Horizon, Organic Valley's market share is quite small.

In addition, urban milksheds with significant numbers of organic consumers also have regional organic companies serving their markets. For example, the San Francisco Bay area has, in addition to Horizon, two local organic dairy companies. One, Straus Family Creamery, is a local Marin County–based dairy composed of two dairy farms. The other, Clover Stornetta Dairy, is a Petaluma-based dairy processor. Both of these companies sell organic milk primarily in the San Francisco Bay area.

The standards versus process approach to organic food certification certainly helps to explain the large-scale character of the organic milk industry. In an organic dairy, one of the main challenges is antibiotics. Dairy cows are susceptible to a particular udder infection known as "mastitis." Cows that are heavy producers, that are in large confined herds, and that are not let out to pasture—in other words, cows kept under industrial conditions—are particularly susceptible to this infection. While these infections can sometimes be kept under control with less drastic methods, a full-blown infection generally requires antibiotic treatment. Most organic dairy standards require at least several months of separation before a cow treated with antibiotics can be returned to the organic herd. In most cases, farmers find it too expensive to keep a treated cow. Therefore, the cow is generally culled from the herd altogether after treatment.

To meet the standard, a farmer can pursue one of two strategies: (1) provide the cow with an environment that prevents the disease, generally residence in a smaller herd with access to outdoor space and pasture, while "pushing" the cow less to produce more milk, or (2) confine cows in a barn, which allows for larger herds, and treat the dairy farm as a quarantine system in which cows are milked until infection occurs, with a rapid turnover of cows.[43]

Does the consumer care whether a dairy producer pursues strategy 1 or 2 to produce organic milk? From the Marxian structuralist perspective, either the consumer is fooled by dairy producers claiming to be philosophically committed to organic while pursuing a quarantine strategy, or the consumer is unconcerned about the actual organic food production process as long as it functions as a display of cultural capital. Is the role of the consumer in the food system really this simple?

The rest of this chapter will argue that the organic milk consumer is more than a dupe of marketing strategy. The consumer is, in fact, a political actor, although an actor in a different form of politics. New cultural approaches to consumption have attempted to reformulate the role of the consumer in the current capitalist economy.[44] From this perspective, "personal choices have political ramifications, and decisions about food can give voice to political commitments."[45]

Organic milk provides a perfect case study for reexamining the role of the consumer in the politics of organic food. Organic milk, as mentioned above, was not part of the original organic food movement, and many of its consumers are not members of any social movement around food. In addition, the presence of large-scale corporate actors in the industry (megacorporations, oligopoly, mergered firms, Wall Street, multinationals, and industrial farms) makes it an excellent candidate for being the "poster child" of the idea that organic agriculture is basically big business in disguise. In other words, if we can find consumer politics here, we can find it anywhere.

To make the claim that consumers have a politics requires defining "the political" as a concept. The most common definition comes from Max Weber, who defined politics in terms of authority, particularly the authority to get others to do what you want them to do.[46] More radical or Marxian definitions emphasize those human choices or activities that transform the capitalist economy and overcome the power of the economic elite. More recently, new approaches to politics have focused on process rather than the challenging of power holders. From this point of view, politics is not only the power to influence others but also "the capacity to act," the ability to envision change and implement that change.[47] This approach focuses on how people envision or "represent" society and how they struggle to implement these visions. Often, from this approach, one examines conflicting media representations and public discourse that advocate for a particular vision of the world.[48] Representation is a form of political action in itself, but it also "represents" real conflict occurring in the material world and provides empirically available evidence that one can "read" to better understand the political struggle that is taking place.

NOT IN MY BODY AS REFLEXIVE CONSUMPTION

To fully explain the role of the consumer in organic food, we need to provide a framework that makes the consumer into a more complex, even human, being. One way to do this is to borrow from other sociologies, particularly the sociologies of science, of the body, and of risk, that have in recent years treated people as actors who participate in the construction of society. A key component in this constructivist approach is social discourse, the process in which various actors make claims, create representations of the world around them, and contest the claims and representations of other actors. People consume in social "communities of practice" interactions that embed actors in everyday relationships.[49] Envisioning the consumer as having an active part in the creation of the food system requires rethinking the nature of political action. A consumer who is not a member of a social movement can still act politically if she or he takes into account various political claims about a product in the process of making a purchase. These claims take place in the public sphere but also in a consumer's community of practice. I will call this process of taking in claims but not necessarily espousing any of them "reflexive consumption."[50]

A reflexive consumer is therefore not a social activist, nor is he or she necessarily committed to a particular political point of view, as espoused by other actors in the public sphere. The reflexive consumer does not necessarily subscribe to the ideologies of new social movements around food, and may evince characteristics of what Marxian approaches would identify as "false consciousness"—a tendency to be swayed by advertising, fads, status purchases, and so forth. However, the reflexive consumer listens to and evaluates claims made by groups organized around a particular food issue, such as genetically modified (GM) foods, and evaluates his or her own activities based on what he or she feels is the legitimacy of these claims.

Reflexive consumers listen to, and sometimes believe, the claims of activist organic food groups. However, organic food activists are only one part of the networks in which consumers live and purchase. These groups produce only one part of the dialogue of claims of that network. Reflexive consumers also pay attention to the mainstream media, other public and private experts, the medical and alternative medicine establishments (including their own doctors and chiropractors), and—perhaps most importantly—their personal networks of friends and relatives. These discursive networks form a part of the "community of practice" in which a person performs acts of consumption.

REFLEXIVITY AS REJECTION: NIMBY AND NIMB

Food is a particularly important focus for reflexive consumers, since food consumption is a negotiation about what a person will and will not let into his or her body. It is a question of "Will I, or will I not, open my mouth to this food?" In this way, as Allen and Sachs have noted, the social movement most closely parallel to the sustainable food movement—and, by extension, the anti-GM food movement—is the "Not in My Backyard" (NIMBY) movement (or "syndrome," depending on your perspective).[51] The NIMBY movement also begins with a refusal—to not allow toxic facilities into a resident's neighborhood. In parallel, the anti-biotech/GM refusal politics could be characterized as a "Not in my Body" or NIMB form of politics.

Comparing NIMBY and NIMB

The most important difference between NIMBY and NIMB is the fact that NIMBY fits the traditional definition of a social movement, while NIMB does not. NIMBY involves a collective refusal to have industrial *facilities* located in a neighborhood, while NIMB involves a controversy over the individual consumption of an industrial *product*. With adequate amounts of information—such as product labeling—NIMB could be carried out on a voluntaristic, individualist basis, whereas local facilities, once located, affect all local residents, involuntarily.

Yet, while NIMB cannot be considered a social movement in the same sense as NIMBY, the following discussion of their similarities indicates that they share "forms of politics." Some of these forms are (1) a contestation of knowledge claims made by economically powerful actors and their experts, (2) attempts at enrollment of publics on one side or another of the issue, and the attendant threats to legitimacy when such enrollment is not successful, and (3) a risk politics that determines who will bear the brunt of possible risk burdens. The following discussion will look at each of these in turn.

Contested Knowledge

In both NIMBY and NIMB cases, business, government experts, and scientists are in conflict with public intuitions about the "truth" concerning the safety of these facilities/products. In both cases, there are significant differences between "scientific" characterizations of risk and "popular epidemiology," the cultural perceptions of risk.[52] (The section on risk politics, below, will look more closely at the source of these cultural perceptions of risk.)

The rBGH controversy involves a widespread rejection of milk produced using this substance, despite a lack of scientific consensus that rBGH-produced milk is unhealthy. Without downplaying the problems of increased insulin-like growth factor (IGF-1) in milk produced using rBGH, the scientific evidence on the risk of this substance—and GM food in general—is controversial compared to well-known, everyday toxic food exposures. Risk claims about rBGH are made mostly by people outside the scientific establishment and by citizen publicizers such as Robert Cohen.

A comparison of rBGH risk claims with toxic exposure related to the consumption (and production) of strawberries provides a good counterpoint. The health and environmental risks of conventional strawberry production and consumption have broad scientific agreement, based on evidence that is widely accepted by the established community of environmental scientists and toxicologists. Many of the substances used to produce strawberries are listed as toxic by the Environmental Protection Agency. Nonorganic strawberries are produced with a variety of pesticides such as Captan. One in four strawberries has Captan residues on the fruit, along with a number of other pesticides. The EPA lists Captan as a "probable human carcinogen" with a "B2" rating, meaning that the animal evidence is sufficient to indicate that a substance is carcinogenic.[53] Such a rating means that you could get a significant proportion of scientists in a room to agree on the risks of Captan. Compared to the controversial evidence around IGF-1, the risk science around strawberries is strongly "institutionalized": the substances consumers are potentially exposed to have established high-risk labels imposed by public institutions, reflecting a general consensus in the scientific community.

In addition, nonorganic strawberries are universally grown using a soil fumigant, methyl bromide. This chemical has been classified as an ozone-depleting substance in the Montreal Protocol, an international environmental treaty. It is also a "Category I acute toxin," which is the highest toxic category defined by the EPA. The anti-rBGH movement paints the dairy industry as forcing a whitewash of the problems with IGF-1 and antibiotics in milk. There is a small amount of inconclusive proof that politicking got rBGH approved by the FDA.[54] In contrast, the series of delays on the ban of methyl bromide use, both internationally and in California, has been publicly proclaimed as political by the mainstream media. This type of special political maneuver took place at a time when many other industries, including the large, politically powerful chemical industry, were successfully phasing out their use of ozone-depleting products.[55]

Despite the general mainstream consensus about the toxicity of, and potential human exposure to, the substances used to produce strawberries, the growth of the organic strawberry industry is not a "hot button" for industry analysts today. From a rational point of view, parents who feed their children organic milk with conventionally grown strawberries are not making the right decision—spending one's scarce dollars on organic strawberries and conventionally produced milk makes more sense. The complaint that people do not make the "right" decisions about risk pervades the NIMBY literature. In both cases, people are making decisions about risks based on other knowledges besides the numbers.

Both NIMBY and NIMB intrinsically challenge the acceptance of risk. Nevertheless, there has been little work on understanding the risk politics around food compared to the enormous literature on NIMBY. A great deal of sophisticated theoretical grappling has occurred in attempts to understand public NIMBY reactions. The lack of theoretical attention to the NIMB public is in part due to its more recent arrival as a "problem" in the social landscape, with the expanding resistance to genetically modified foods. But in part this has also been due to the lack of attention to the consumer in food system studies. For this reason, people trying to understand NIMB would do well to look at the conversations that have taken place in the attempt to understand NIMBY.

Enrollment

Both NIMBY and NIMB challenge generally larger, often multinational, corporations to convince the public that their facility/product is "safe." NIMBY traditionally, and NIMB increasingly, involve increasingly broad enrollment of actors—various publics, government experts, university and private scientists—in broad, policy-related, civic discourses about what to "do" about these facilities/products. Neighborhood industrial facilities and food share the characteristic of being "local." They both find us where we live. This proximity is magnified when the product in question is a (northern European) cultural staple, like milk.

In both NIMBY and NIMB, the "actor" is not simply the activist. While the social movements prompting the initial refusal are important—even necessary—the *reaction* to that refusal brings everyone "to the table," that is, into the field or network of discourse. The owners of a potential facility under attack by NIMBY activists do not respond simply to the activists, they address their communicative strategies to all local residents (Do *all* of you really not want this facility? Are *all* of you really refusing these jobs? Do you really believe these ill-informed extremists?). Politicians (state and local), the activists themselves, and other local groups attempt to enroll the average resident into the discussion (Do you really want to be poisoned? Do you really believe these big businessmen and the government officials in their pockets?). In the same way, those actors under challenge in the GM food discourse—biotech companies, government, agricultural research universities, and research institutions such as the Rockefeller Foundation—do not simply respond to activist groups. Instead, they attempt to enroll regular consumers and taxpayers to their side, or at least to neutrality on the subject. Leaders of these institutions know that their legitimacy is at stake, but that the average consumer will not necessarily challenge them directly on this issue. Instead, if the active enrollment of the mainstream public does not succeed, biotech companies, governments, institutions, and others suddenly find themselves with mysterious, untraceable legitimacy "leaks," which result in a lower stock price, a strong election challenge, or a lack of critical support in other areas.

Some stock analysts valued Monsanto's stock price below zero after several allegations that its genetically modified (GM) crops were harmful to people and the environment.[56] The rejection of GM foods in Europe is one example of the powerful force of the reflexive consumer. GM food has not gotten the same kind of political attention in the United States; however, U.S. stockholders participated in the large-scale investor withdrawal of support for Monsanto, after European governments questioned the safety of biotech products. Stock price showed remarkable sensitivity to the reflexive European consumer. While U.S. consumers remained suspicious of homegrown anti-GM activists such as Jeremy Rifkin, Europeans and European governments relied heavily on Rifkin's analysis to question these products.[57]

Risk Politics

There is a significant social science literature on the politics of risk, much inspired by attempts to understand the NIMBY phenomenon.[58] One of the most important findings of these studies is that people rank a risk according to their ability to control that risk, that is, the extent to which the risk is voluntary.[59] On first glance, it would appear that the analyses of risk politics do not help us understand the voluntary purchase of organic milk. Food, like cigarettes, is a voluntary risk. However, when the food is a cultural staple—that is, deemed necessary by the predominant culture—control becomes a very important factor. If consumers believe that they, and especially their children, *have to* drink milk, then the potential unwanted exposure is nearly as involuntary as the siting of a local toxic facility.

Therefore, while the individual consumer is not a political activist, his or her consuming actions are embedded in a larger network of politics that includes the state, science, and social movements as well as the gendered politics of the family (as in who decides what's in the refrigerator). Like the risk politics of NIMBY, NIMB politics involves issues of control. For consumers, part of that control includes purchase, and part of that politics includes demands for the provision of information—such as product labeling—on which to make consumption decisions. Social activists also attempt to enroll reflexive consumers to perform political activities such as boycotts.

From this point of view, the consumer not only buys products but thinks about consumption both as an individual and as a member of a community of practice. As part of this network, the consumer will not only buy products, but will also engage in a public and private dialogue with other consumers, political activists, government experts, scientists, and others about the consumption decisions he or she is making. Therefore, even the purchases of consumers who are not overtly involved in food politics may contain a "form of politics."

The concept of reflexive consumption provides a lens that enables us to see the politics in consumption activities. In social constructivist fashion, reflexive consumption moves our search for power away from the consumer's consciousness to the discourse surrounding the consumer in the world in which she or he acts. Interestingly, organic milk companies have initiated a new form of discourse with their consumers: the milk carton itself. For many years the carton was reserved primarily for pictures of missing children, but organic milk companies have reclaimed that space to represent itself and its product to the consumer (commonly known as "marketing"). In response, many conventional companies are following suit. The following discussion will analyze a few of these "talking" milk cartons, to understand the nature of the political discourse around organic milk.

THE TALKING MILK CARTON

The panels on the organic milk carton "talk" to the consumer. Yet, when we examine the types of talk on organic milk packages, it becomes clear that there are a number of different claims being made and that the organic and non-rBGH milk companies represent the relationship between the producer and the consumer in different ways. These different claims represent the enrollment practices of actors in different positions within the market and within the contested discourse on food. The claims of organic companies on their milk cartons are distinctly different from conventional milk company claims and fall into three major categories: consumer-as-authority, agrarian, or neighborly.

Horizon: The Politics of Consumer-as-Mother/Authority

Horizon's talking milk carton exudes friendliness. "You deserve delicious foods that are safe and healthy," begins one Horizon milk carton. Right away, the consumer is made sovereign in the conversation. "This kind of quality begins right at the farm," the milk carton continues, linking the consumer's desire to the way the company treats nature. "And, by not using pesticides," the carton concludes, "we keep them out of your family's food." The company is telling us it is on our side; it knows what we want and will provide it to us. It knows what we do not want, and will keep it away from us.

The major image on the Horizon milk package is the cute cow, or the "clean living cow," as its milk carton commonly calls her (fig. 10.1). The package's designers recently featured Horizon's new "look" on its Web site. The challenge, the designers state, was that Horizon's original package "wasn't designed with a national audience in mind."[60] But with Horizon's new national market, and increasing emphasis on conventional distribution channels like supermarkets, it now "has main-

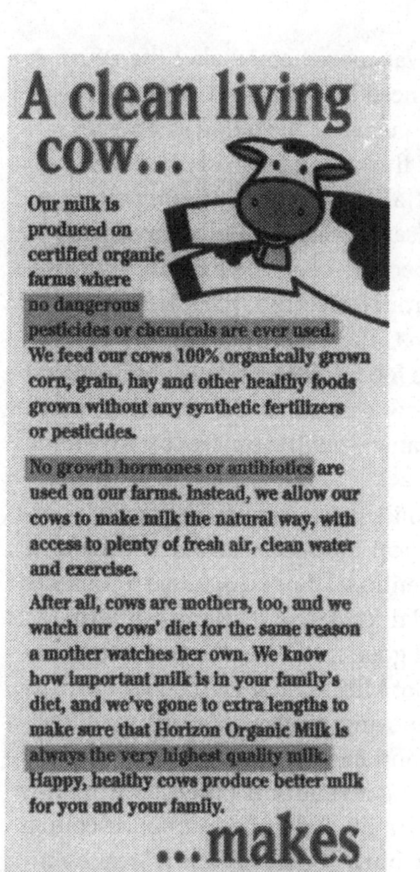

A clean living cow...
Our milk is produced on certified organic farms where no dangerous pesticides or chemicals are ever used. We feed our cows 100% organically grown corn, grain, hay and other healthy foods grown without any synthetic fertilizers or pesticides.
No growth hormones or antibiotics are used on our farms. Instead, we allow our cows to make milk the natural way, with access to plenty of fresh air, clean water and exercise.
After all, cows are mothers, too, and we watch our cows' diet for the same reason a mother watches her own. We know how important milk is in your family's diet, and we've gone to extra lengths to make sure that Horizon Organic Milk is always the very highest quality milk. Happy, healthy cows produce better milk for you and your family.
...makes real good milk.

FIG. 10.1. A Horizon milk carton panel, 2000. © 2000 Horizon Milk Company. Reprinted by permission.

stream customers in addition to its core of health food shoppers. Horizon is shifting its market strategy. The new target buyers of Horizon products are highly educated mothers."[61]

The cute cow is an American cultural phenomenon in itself, associated with the "soft" side of social provision (the family) as opposed to the "hard" side (capitalism). The *Far Side* cow jokes, the ubiquitous cow coffee cups, the Holstein-dotted Gateway Computer packaging, the cow statues grazing on Chicago streets—all of these cute cows represent a friendly, controllable, yet natural provision system, a sort of identity-based pastoral ideal. As described in chapter 5, cow images disappeared from milk advertising around the turn of the century. The major exception was Borden's Elsie, who was more homemaker than cow. A recent Elsie nostalgia craze, as evidenced by the amount of Elsie memorabilia for sale on e-Bay, reflects the current iconization of bovines.

The Horizon carton tells us how this cow lives, without pesticides, hormones, or antibiotics, because, "[a]fter all, cows are mothers, too, and we watch our cows' diet for the same reason a mother watches her own." The cow and the consumer become one. The carton portrays a cute cow flying through the air. Far from the large, crunching behemoth one commonly finds in dairy barns, the cute cow emphasizes the friendly, nonthreatening status of the food production system responsible for the milk in the carton. The cow represents the harmlessness of the milk—the quarantine side of organic—not the process by which the milk is made.

The uniqueness of the Horizon milk discourse becomes most obvious when compared to the more typical producer-consumer discourse around conventional milk. The conventional dairy discourse represents the quality of its milk according to claims of expertise and authority, as in, "We are the ones who know what good milk is, and you are the beneficiaries of this knowledge." The "Got Milk?" campaign plays with this authoritarian discourse, with its imposing question and its mass-produced milk mustache on celebrities and figures of authority.

Berkeley Farms, for example, is a conventional dairy that communicates to its consumer through the traditional authoritarian discourse (fig. 10.2). Its image is not a cute cow but a "Seal of Quality," representing a certification by a governmental authority. The conversation on the carton reflects this authority discourse, talking not about cows or nature, but about field managers as experts who inspect dairy farms. The carton is telling us that we should trust the milk because experts are employed in its production. The milk is also superior, according to the ad, because it exceeds state quality requirements. Finally, the ad announces that its independent dairy farmers are "nutritional ecologists" who provide "skilled husbandry."

In fact, the Berkeley Farms carton gives a great deal more information about milk production than the Horizon Organic ad. It makes a real commitment to extra inspection personnel, and it tells the consumer that by the standard measures of bacterial contamination, it is very clean milk. It tells us where the farms are. However, compared to the "deserving" mother/authority consumer of the Horizon Organic ad, the Berkeley Farms carton misses the mark in the current post-Fordist

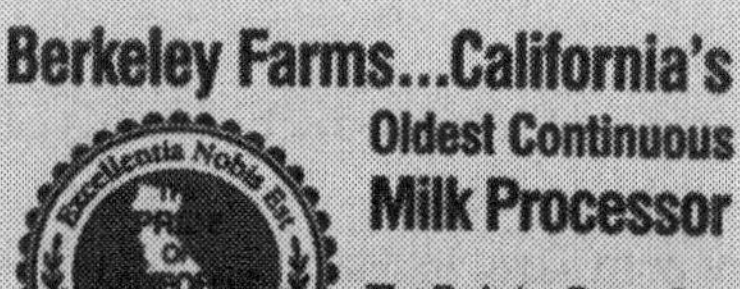

Berkeley Farms...California's
Oldest Continuous
Milk Processor
Excellentia Nobis Est
BERKELEY FARMS
The Berkeley Farms Seal of Quality is a symbol of pride, assuring consumers of the dairy's commitment to health and wellness, since 1910.
Field Management Makes a Big Difference
Berkeley Farms is the only Bay Area milk processor employing a field manager whose sole job is to inspect dairy farms that ship milk directly to Berkeley Farms' milk processing plant. The field manager guarantees that Berkeley Farms' milk exceeds state quality requirements for both standard plate and somatic cell counts.
Response to Consumers Makes a Big Difference
Berkeley Farms is sensitive to prevailing consumer concerns over rBGH, the synthetic bovine growth hormone. Although federal tests prove that there is no significant difference between rBGH treated milk and non-rBGH treated milk, Berkeley Farms continues to ban the use of rBGH and enjoys the support of its independent dairy farmers who certify rBGH free milk.
Family Farms Make a Big Difference
Carefully chosen family farms provide Berkeley Farms milk from the rolling hills of Marin and Sonoma Counties and the lush San Joaquin Valley. Berkeley Farms and its independent dairy farms consider themselves "nutritional ecologists," respecting the environment and providing advanced, skilled husbandry throughout the milk production process.
THE PRIDE OF CALIFORNIA
SINCE 1910

FIG. 10.2. A Berkeley Farms milk carton panel, 2000. © 2000 Dean Foods. Reprinted by permission.

milk discourse. The consumer wants control over the product, not assurance that other authoritative people are controlling the product for them. Of course, the idea of consumer control is a myth, but the cute cow masks the human system—the producers, the experts, the government agents—necessary to make the cow live "clean."

The carton talk of other conventional milk companies takes authority away from the consumer in another way. Many conventional milk cartons advise the consumer to drink milk to get enough calcium. Here the consumer becomes the recipient of nutritional advice. The Horizon carton, on the other hand, makes no claims of nutritional authority. The consumer "knows" what is good and chooses Horizon's milk because of that knowledge.

Straus and Organic Valley: Agrarian

Not all organic milk companies represent their product in terms of cute cows. For example, the bottle talk provided by Straus Family Creamery, the small, local organic dairy company in Marshall, California, emphasizes agrarian issues. Straus's glass bottles provide less room for talk, but the difference in message is clear. Buying Straus milk enables a family farm to continue in agriculture. "Dairy farms are disappearing at a rate of about 5% a year," states one bottle. "Going organic gave our family the chance to continue farming." Another bottle gives a history of the dairy and ends with "Thanks for your support." The consumer in this bottle talk is a small farm supporter. Both farm and consumer, in this talk, are pulling together for a particular vision of agriculture.

Like Straus, Organic Valley, the cooperative organic milk company, uses farm images on its cartons to represent its product. Cows appear on the carton, but they are not anthropomorphic cute mom cows. The Organic Valley carton shows a pastoral scene, with a more naturalized (less human) cow in association with the farm family. This portrayal is reminiscent of the romantic pastoral images of the mid-nineteenth century, in which the tending milkmaid represented the care of nature. The Straus and Organic Valley dairy discourses emphasize agrarian values and cooperation between farmer and consumer, and farmer and nature, for a particular way of life. Yet, unlike the earlier "The Farmer Feeds Us All" ideology of agrarian producerism, the new agrarian message is, "We can't survive without you."

Clover Stornetta: Neighborly

Clover Stornetta is a regional Bay Area milk company that sells both organic and non-rBGH milk. Although this company buys milk from a group of local farms around Petaluma, Clover Stornetta does not emphasize the agrarian nature of its product. Instead, like Horizon, the company emblazons its milk cartons with a cute cow. The milk talk on the carton, however, does not address the consumer as sovereign, but as a neighbor. The cartons often announce local community events around the Bay Area, serving more as a community billboard than as a mouthpiece for the company. This emphasizes the localness of the company, compared to Horizon, which cannot legitimately make the same neighborly claims.

REFLEXIVE CONSUMPTION AS A LESS PERFECT FORM OF POLITICS

How does the consumer actually respond to these claims? Is she or he an authority, a neighbor, an agrarian? Not necessarily. The reflexive consumer is an actor in a larger network that involves more than simply reading a milk carton. Is the consumer simply a victim of false consciousness, in which organic dairy companies fool them with visions of happy cows, saved farms, and neighborly communities? Not necessarily. Consumers evaluate claims and act on these claims every time they reach for a milk carton or bottle at the store. For example, there is a growing awareness in the food activist community that Horizon's post-Fordist organic milk production strategy is incompatible with the political philosophy around organic food. As a result, there is a growing challenge to the way Horizon makes its organic milk. Recently, a newsletter by an upstate New York consumer-farmer coalition, the Regional Farm and Food Project, criticized the way Horizon produced its organic milk.[62] Rural Vermont, a rural activist organization, also criticized the relationship between Horizon and Suiza. The article advised consumers, "Don't assume organic means unplugged [from the global food system] . . . Horizon has two massive factory farms, one in Idaho and the other in Maryland, that supply a large proportion of its milk."[63] A national consumer food safety group, the Pure Food Campaign, in its electronic newsletter article titled "Organic Standards: Who Really Speaks for the Organic Consumer?" distinguishes between the "rank-and-file" organic movement and the "'Big Players' in the natural foods industry."[64] The newsletter accuses the big players of lobbying only for those parts of the USDA's proposed national organic rules that would maintain a separate identity for organic, but not preclude industrial forms of organic production.

It remains to be seen, however, what reflexive consumers will do once (and if) they are confronted with a choice between Horizon's quarantine organic milk production strategy and a smaller-scale, process-based alternative. Will consumers care one way or another? On a larger scale, what will consumers choose if the GM food industry actually creates a cheaper, more convenient, more nutritious, more delicious strawberry, maybe even one that can be grown without methyl bromide? What if that bomb of the GM industry, the Flavr-Savr tomato, did actually taste good for several days longer than a regular tomato?

Neither the progress nor the downfall stories help us answer these questions. Reflexivity as a form of social action requires that we look at the world in a new way, one that doesn't start with a single, fixed idea of perfection. To redeem our roles as political actors, we must start by accepting the notion of an imperfect world. Only then can we make the world better.

"Imperfect" in this case doesn't simply mean faulty, or broken. In an imperfect world we need to gain an awareness of the bargains we strike—with nature, with economic forces, and with each other—when we forge the future. This does not mean we are denying the power that economic forces have over us. But those forces benefit one small group of people at the expense of the rest. Those people legitimize their privileged positions by telling a story that glorifies the processes that put them on top. This story, the perfect story, veils the unequal and unjust effects of industrialization, postindustrialization, and globalization by making questions about the people behind the power unaskable. Imperfect stories enable us to bring these questions to the fore, to make them the center of the conversation.

However, reflexivity is not a story in itself. It is a tool to enable us to create new stories, better imperfect stories, about food and about society as a whole. Imperfect stories enable us to understand ourselves and our world the way perfect stories never can.

Afterword

Today, many Americans mourn the loss of life's diversity and fear the creeping specter of sameness that modern life can bring. In the ultimate downfall story, our future society will be a place where successive generations of designer babies look more and more alike; where we sit in exactly the same position as we work in front of the computer screen; where soybeans, corn, cotton, and other agricultural crops each come from genetically identical seeds and all those genetic varieties now rotting in public seed banks will be gone forever. The only variety will be the vast number of ways businesses exploit our longing for variation, producing endless stylistic changes for us on a mass scale.

The sociologist Max Weber, looking at the rise of modern society, predicted this nightmare scenario. Using the army as an example, he compared the rationalized modern army to the individual heroism of the heraldic past. In the feudal days of hand-to-hand combat, each knight fought in his own way and each excelled at a particular way of fighting. The modern army, however, succeeded only when everyone did the same thing, as a unit. Soldiers needed to kneel, aim, and shoot in disciplined infantry lines. Unique strategies and individual approaches led to certain defeat against this highly disciplined fighting machine.

The idea of perfection made efficiency a moral imperative. Yet, while it seemingly led us to a better society, this idea in fact provided the intellectual basis for the creation of sameness. Efficiency became both a welcome new future and an inevitable iron cage.

Yet the story of milk shows us that we are the masterminds of the cage that seeks to shut us in. We build the cage every day, with our stories. For Weber, once the moral ideas of Calvinist Protestantism (work for the sake of work) set the stage, the machine of modernity (capitalism, bureaucracy, rationalization) took on a life of its own. No amount of counter-thinking, except perhaps the temporary following of a charismatic leader, could overcome this force. The story of milk, however, shows us the way ideas about modernity, purity, and perfection must be continually renewed to keep the machine of modernity in working order. Ideas about perfection have changed over time, and these ideas were more than simple reflections of economic changes; they were integral to the way people made their future. Without a moral framework of perfection in which to stand, the modern economy cannot continue to exist.

The ideal of perfection—that there is one best way of living and that this way is the best in all circumstances—both feeds and is fed by the juggernaut of rational efficiency. Without this ideal, the whole system comes to a halt. As consumers, we demand perfection and efficiency: the product produced cheaply but without fault. We choose cheese "food" rather than risk buying a cheddar cheese that may not be as delicious as the last. While singing Joni Mitchell's "give me spots on my apples, but leave me the birds and the bees," we pick the spotless apples out of the supermarket bin. We fight the signs of aging—evidence of a decline from perfection. In all these cases, our pursuit of a faultless, universal, efficient, and inexpensive existence drives the forces of rationalization forward.

At least, on an assembly line, it was possible to change one's endless clock-like motion for another clock-like motion down the line (to car hood rather than car door assembly, for example). Now, the person in the next computer station is in the exact same physical position, performing the same physical action. Only the mindwork changes.

Once, in a small grocery in France, a French companion gleefully picked up a small round of Camembert cheese and put it under my nose. "Ça pu comme les pieds" (It smells like feet), he said proudly. The best cheese, in his estimation, smelled the worst. The best part of this food was the fact that it didn't smell like roses. It had a fault, a wonderful fault. In the same way, other friends of mine wax poetic about the heavenly taste of durian, a Pacific Island fruit that everyone agrees smells distinctly like fecal matter.

The modern industrial system could never have created Camembert or durian. These products were faulty, and in the industrial, rationalized system of "quality control," such characteristics would be the first to go. Of course, there have been exceptions to this tendency, one of which was a particular kind of Limburger cheese, made by Kraft, a leftover from a time when Limburger was made in upstate New York to satisfy the German immigrant population. I would serve this delicious cheese—which did not smell exactly like feet, but close—to anyone who could stand it. It gained converts and soon a small coterie of friends were searching the shelves for that small square in the old-fashioned silver-foil wrapper.

Then one day it was gone, replaced by a jar filled with a cream-cheesy-like substance labeled "Limburger cheese food." This quirky little Kraft product had gone the way of the passenger pigeon, the subsistence farm, and the summer hill dairy. Certainly the disappearance of a package of Limburger from the grocery shelves is not an important event. Whole ways of being and ways of life have disappeared or are under threat. But food is life, written small and daily. We can learn from it.

We insist on perfect food in large part for our children. We don't want them at risk. Yet think of the risk our children face in a world controlled by ideas of perfection. Whether they are able to fit into these ideals—of race, gender, intelligence, beauty—or not, they will lose. They will lose either through a loss of individuality or through the marginalization of not fitting in.

Can we change this? Only if we recognize the fragility of the framework of stories this whole system rests on. We can reject perfection and begin to forge less than perfect lives and a less than perfect politics. We will then recognize that the system is not simply an autonomous force that we must resist. Instead, we need to focus on creating new stories and the new communities of practice that go with these stories. If we do this together, the system will have to change with us.

To do this will require understanding that politics does not happen only when people take to the streets. We make political decisions simply by living our everyday lives. Every meal is a political act. So is everything we buy, every conversation we have, and how we show affection or anger. This everyday politics appears invisible to most because we haven't found a way to talk about it. But as we articulate the politics of everyday life, we will increase our ability to respond to this activity in a reflexive way.

In particular, we need to reflect on the many stories of perfection, not just about milk, but about all aspects of our lives and our society. We need to find and reflect upon the contradictions, the compromises, the bargains, that is, the imperfections that are hidden behind stories of perfection.

How do we learn to tell other stories about who we are and where we are going? First, we need to listen to the stories of people who never fit into the perfect stories, for example, those people for whom milk was never the perfect food. Second, we need to reject the story of perfect nature, either as God's design or as a social goal of progress. Nature never had a perfect food to give us. Let's cherish the wonderful imperfections—including milk—she has to offer.

Notes

NOTES TO CHAPTER 1

1. For example, McMahon, in her study of food in widows' inventories in New England, found that "the guarantee of at least one milch cow was the most common right granted to a widow in her husband's will." McMahon, "A Comfortable Subsistence: A History of Diet in New England, 1630–1850."

2. Bidwell and Falconer, *History of Agriculture in the Northern United States;* Giblin, *Milk: The Fight for Purity;* McIntosh, *American Food Habits in Historical Perspective;* Danbom, *Born in the Country.*

3. Shaftel, "A History of the Purification of Milk in New York," 275.

4. Atkins, "White Poison? The Social Consequences of Milk Consumption, 1850–1930."

5. Gates, *The Farmer's Age: Agriculture, 1815–1860.*

6. The earliest New York City study of breast feeding rates, in the 1910s, found that three-quarters of all infants were breast fed. However, the study also found that 75 percent of all infants who died of enteritis were fed on cow's milk or patent foods. Shaftel, "A History of the Purification of Milk."

7. Spencer and Blanford, *An Economic History of Milk Marketing and Pricing,* 352. This amount includes the milk equivalent of cream.

8. For those interested in my specific work on rBGH, see Geisler and DuPuis, "From Green Revolution to Gene Revolution"; DuPuis and Geisler, "Biotechnology and the Small Farm"; and DuPuis, "No Limits to Growth."

9. Friedland et al., *Manufacturing Green Gold: Capital, Labor and Technology in the Lettuce Industry.*

10. Chandler, *The Visible Hand: The Managerial Revolution in American Business.*

11. Benjamin, "Theses on the Philosophy of History," 261, 255.

12. Melucci, *Nomads of the Present.*

13. Here I am following the ideas of Michel Foucault as described in *The History of Sexuality.*

14. These ideas borrow from deconstruction as a method of reading history, although many deconstructionists would critique my optimism about finding the truly social behind the veil of modern texts.

15. This approach has been inspired by the writings of Stuart Hall, particularly his concept of "the politics of representation" as articulated in his essay, "New Ethnicities."

16. In this case, I am using a conception of politics first introduced by Antonio Gramsci in his *Prison Notebooks*.

NOTES TO CHAPTER 2

1. Spann, *The New Metropolis*.

2. Spann, *The New Metropolis*, 121.

3. Hartley, *An Historical, Scientific and Practical Essay on Milk*, 139.

4. Hartley, *Essay on Milk*, 139.

5. While Prout's *Chemistry, Meteorology, and the Function of Digestion* may not be the first statement concerning the perfection of milk, Hartley's substantial reference to this work shows that it was clearly one of the most influential.

6. Infant mortality rose from 32 percent in 1814 to 50 percent in 1841. Shaftel, "A History of the Purification of Milk in New York."

7. New York City infant death rates cited in Meckel, *Save the Babies*, based on Griscom, *The Sanitary Condition of the Laboring Population*. Term "infant abattoirs" used by Meckel. Discussion of rural versus city infant death rates in Preston and Haines, *Fatal Years*.

8. Randall's Island death rates from Meckel, *Save the Babies*. Quote from Meckel, 45.

9. Apple notes the almost universal recommendation by physicians that a mother breast feed her children. Apple, *Mothers and Medicine*. Preston and Haines provide an overview of public health physicians' views on the topic. Preston and Haines, *Fatal Years*.

10. Many upper-class families moved out of the city altogether during the summer, in order to escape what was obviously an unhealthy atmosphere.

11. Systematic collection of infant mortality statistics did not begin until the second half of the nineteenth century. For an overview of infant mortality rates in the late nineteenth century, see Preston and Haines, *Fatal Years*; and Mullaly, *The Milk Trade of New York and Vicinity*.

12. Flexner, "The Battle for Pure Milk in New York City," 161.

13. For biographical information on Hartley, see Hartley, *Memorial of Robert Milham Hartley*; Rosenberg and Smith-Rosenberg, "Pietism and the Origins of the American Public Health Movement"; Smith-Rosenberg, *Religion and the Rise of the American City*; Walters, *American Reformers*; Thomas, "Romantic Reform in

America"; and Schlossman, "The 'Culture of Poverty' in Ante-Bellum Social Thought."

14. Quoted in Thomas, "Romantic Reform in America," 667.

15. Thomas, "Romantic Reform in America"; Smith-Rosenberg, *Religion and the Rise of the American City.*

16. Hartley, *Essay on Milk.*

17. Historians differ widely on the question of what motivated antebellum evangelicals to become active in social reform. Joseph R. Gusfield argued that the reformers were displaced elites attempting to reinstate their status through new forms of social control. Gusfield, *Symbolic Crusade: Status Politics and the American Temperance Movement.* A number of authors have argued against this perspective, noting that the majority of moral reformers were from the new elite of urban entrepreneurs and that "in some guise or another, evangelical Protestantism was the religion of most Americans." Walters, *American Reformers.* More recent approaches describe how reformers, though a minority of the American population, drew from widely shared dominant beliefs about evil, perfection, and the millennium. See Davis, *Ante-Bellum Reform*; Whiteaker, *Seduction, Prostitution, and Moral Reform in New York, 1830–1860*; Walters, *American Reformers.* From a milk reform perspective, what is more important is the ultimate effect of their action in building a reform lobby and how this lobby made public claims about milk.

18. Cross, *The Burned-Over District.*

19. Schlossman, "The 'Culture of Poverty' in Ante-Bellum Social Thought."

20. Finney quoted in Ahlstrom, *A Religious History of the American People.*

21. Pegram, *Battling Demon Rum,* 18.

22. Pegram, *Battling Demon Rum*; Whiteaker, *Seduction, Prostitution, and Moral Reform in New York.* For a history of the extent and impact of the Second Great Awakening on social life and social reform movements in the northeastern states, see Cross, *The Burned-Over District,* and Barkun, *Crucible of the Millennium.*

23. Smith-Rosenberg, *Religion and the Rise of the American City.*

24. Whiteaker, *Seduction, Prostitution, and Moral Reform in New York*; Schlossman, "The 'Culture of Poverty' in Ante-Bellum Social Thought."

25. Whiteaker, *Seduction, Prostitution, and Moral Reform in New York.*

26. Hartley, *Essay on Milk,* 34.

27. Josephus Book 1: 9, quoted in Hartley, *Essay on Milk,* 37.

28. Hartley, *Essay on Milk,* 37.

29. Hartley, *Essay on Milk,* 39.

30. Hartley, *Essay on Milk,* 27.

31. Mullaly, *The Milk Trade of New York and Vicinity,* 110.

32. Simoons, "The Determinants of Dairying and Milk Use in the Old World"; Bryant et al., *The Cultural Feast,* 36.

33. Carper, *Milk Is Not for Everybody;* Johnson et al., "Lactose Malabsorption."

34. Giblin, *Milk: The Fight for Purity,* 9.

35. Interview with Dr. William Greathouse, coroner for Jefferson County, KY. Video on CNN "American Presidents" Web site (http://www.american-presidents.org/presidents/president.asp?PresidentNumber=12).

36. Giblin, *Milk: The Fight for Purity,* 7.

37. Pirtle, *History of the Dairy Industry,* 18; Schmidt and Ross, *Readings in the Economic History of American Agriculture,* 407.

38. Atack and Bateman, *To Their Own Soil.*

39. Lysaght, introduction to *Milk and Milk Products.*

40. Van Winter, "The Consumption of Dairy Products in the Netherlands," 3.

41. Gibson and Smout, "From Meat to Meal." They note that "On milk the evidence is uncertain and historians divided" (18), but that "[o]n the whole the balance of evidence for much of the eighteenth century seems to favor Hamilton [a historian who argues that little milk was consumed]." They cite Hamilton's statement that "Dairy farming was as yet little practised, cows being valued for their powers of draft and for their meat" (18–19).

42. For Sweden (and quote), see Salomonsson, "Milk and Folk Belief"; for Germany, see Teuteberg, "The Beginnings of the Modern Milk Age"; for Scotland, see Gibson and Smout, "From Meat to Meal," 18, and Fenton, "Milk Products in the Everyday Diet of Scotland"; for Hungary, see Kisbán, "Milky Ways on Milk-Days."

43. See note 33; for U.S., see Bidwell and Falconer, *History of Agriculture in the Northern United States.*

44. Rush, *An Inquiry into the Effects of Ardent Spirits.*

45. Taylor, *Eating, Drinking, and Visiting in the South.*

46. Spencer and Blanford, *An Economic History of Milk Marketing and Pricing;* McIntosh, *American Food Habits in Historical Perspective.* Nineteenth-century statistics on milk consumption are difficult to ascertain. Statistics are derived either (1) by estimating the average milk production per cow, multiplying by the number of cows, and then dividing by the human population, or (2) by examining milk supplies delivered to cities. The first form of estimation depends heavily on an accurate assessment of average milk yield per cow; the second relies on milk transport records, which underestimate both rural consumption and urban production in swill dairies. For a discussion of the data problems of dairy consumption statistics, see Spencer and Parker, *Consumption of Milk and Cream.*

47. The quotes are from Hedrick, *A History of Agriculture in the State of New York,* 220, 226, 227. Quote on cider from John Taylor, quoted in Hedrick, 226.

48. Quoted in Hartley, *Essay on Milk,* 3.

49. Ray, *The Wisdom of God;* Newton, *Opticks.*

50. Prout, *Chemistry, Meteorology, and the Function of Digestion,* 260.

51. Prout, *Chemistry, Meteorology, and the Function of Digestion*, 260.
52. Prout, cited in Hartley, *Essay on Milk*, 106.
53. Mullaly, *The Milk Trade of New York and Vicinity*, 116, 115.
54. Hartley, *Essay on Milk*, 334.
55. Flexner, "The Battle for Pure Milk in New York City."
56. Spann, *The New Metropolis*, 121.
57. Hartley, *Essay on Milk*, 319–20.
58. Trall, introduction to Mullaly, *The Milk Trade in New York and Vicinity*, iii.
59. Spencer and Blanford, *An Economic History of Milk Marketing and Pricing*, 354.
60. Spencer and Parker, *Consumption of Milk and Cream*, 25.
61. Jacobsen and Outlaw, "Dairy Product Consumption and Demand."
62. Roadhouse and Henderson, *The Market-Milk Industry*, 608.
63. Roadhouse and Henderson, *The Market-Milk Industry*, 608.
64. Roadhouse and Henderson, *The Market-Milk Industry*, 3.
65. Roadhouse and Henderson, *The Market-Milk Industry*, 6, 7.
66. Roadhouse and Henderson, *The Market-Milk Industry*, 8.
67. Shaftel, "A History of the Purification of Milk in New York," 275.
68. The figure represents consumption as the receipts of milk and cream per capita, which is only an estimate of consumption and may underestimate early consumption because it does not record milk produced within the city.
69. Giblin, *Milk: The Fight for Purity*, 3.
70. For a similar argument, see Campbell, *The Romantic Ethic and the Spirit of Consumerism*.
71. Pirtle, *History of the Dairy Industry*, 18.

NOTES TO CHAPTER 3

1. Hartley, *Essay on Milk*, 205.
2. Hartley, *Essay on Milk*, 205.
3. Subsequent research showed that cow's milk was in fact significantly different from human milk and had to be modified before being adequately digestible by infants. Meckel, *Save the Babies*.
4. By 1883, women in these areas had returned to breast feeding and the rate of death had dropped precipitously. Eager, "Morbidity and Mortality Statistics as Influenced by Milk."
5. Stuart-Macadam, "Breastfeeding in Prehistory," in Stuart-Macadam et al., *Breastfeeding*; Fildes, "The Culture and Biology of Breastfeeding"; Baumslag and Michels, *Milk, Money and Madness*.
6. Preston and Haines, *Fatal Years*; Fildes, "The Culture and Biology of Breastfeeding."

7. Eager, "Morbidity and Mortality Statistics as Influenced by Milk," 230.

8. Salmon, "The Cultural Significance of Breastfeeding and Infant Care."

9. Williams and Galley, "Urban-Rural Differentials in Infant Mortality in Victorian England"; Rosenberg, "Breast-Feeding and Infant Mortality in Norway, 1860–1930"; Liestøl et al., "Breast-Feeding Practice in Norway, 1860–1984"; the classic U.S. urban study is Woodbury, *Infant Mortality and Its Causes.*

10. Fildes, "Breast-Feeding in London, 1905–19," and *Breasts, Bottles and Babies;* Moring, "Motherhood, Milk and Money"; Williams and Galley, "Urban-Rural Differentials in Infant Mortality."

11. Stuart-Macadam, "Breastfeeding in Prehistory"; Fildes, "The Culture and Biology of Breastfeeding," and *Breasts, Bottles and Babies;* Baumslag and Michels, *Milk, Money and Madness.*

12. Hartley, *Essay on Milk,* 203.

13. Hartley, *Essay on Milk,* 96.

14. Hartley, *Essay on Milk,* 97.

15. Brumberg, "The Appetite as Voice."

16. Hartley, *Essay on Milk,* 206.

17. Hartley, *Essay on Milk,* 206.

18. Treckel, "Breast Feeding and Maternal Sexuality in Colonial America."

19. Fildes, "Breast-Feeding in London," and "The Culture and Biology of Breastfeeding."

20. Kintner, "Trends and Regional Differences in Breast Feeding in Germany."

21. Woodbury, *Infant Mortality and Its Causes.*

22. Golden, *A Social History of Wet Nursing in America,* 126.

23. Meckel, *Save the Babies;* Levenstein, *Revolution at the Table.*

24. Apple, *Mothers and Medicine;* Meckel, *Save the Babies;* Fildes, *Breasts, Bottles and Babies.*

25. Eskay's Food was a product of Smith, Kline, and French, which is now part of the major pharmaceutical company SmithKline Beecham.

26. Laird, *Advertising Progress.*

27. An examination of New York City directories from the 1840s to the 1880s uncovered a few milk advertisements, but none mentioned infant feeding.

28. Grant, *Raising Baby by the Book,* 15.

29. Grant, *Raising Baby by the Book,* 20.

30. Allen, *Feminism and Motherhood in Germany.*

31. Golden, *A Social History of Wet Nursing in America;* Fildes, "The Culture and Biology of Breastfeeding."

32. Golden, *A Social History of Wet Nursing in America.*

33. Spencer and Blanford, *An Economic History of Milk Marketing and Pricing.*

34. Golden, *A Social History of Wet Nursing in America.* Better data from the early twentieth century are in Woodbury, *Infant Mortality and Its Causes;* and Fildes, "Breast-Feeding in London," and "The Culture and Biology of Breast-feeding."

35. Ross, *Love and Toil: Motherhood in Outcast London,* 141.

36. Levenstein, *Revolution at the Table.*

37. Rosenberg, "Breast Feeding and Infant Mortality in Norway"; Ross, *Love and Toil.*

38. Brumberg, "The Appetite as Voice."

39. Fildes, *Breasts, Bottles and Babies,* and "Breast-Feeding in London"; Sussman, *Selling Mothers' Milk;* and Kintner, "Trends and Regional Differences in Breast Feeding in Germany" are all examples of studies that found a relationship between artificial feeding and female labor force participation. However, Fildes shows that the rates of breast-feeding were higher among poor women, who were more likely to work. Many contemporary studies show the relationship between the duration (but not the rate) of breast-feeding and mother's employment. For example, see Visness and Kennedy, "Maternal Employment and Breast-Feeding."

40. Fildes, "The Culture and Biology of Breastfeeding."

41. Turbin, "Beyond Conventional Wisdom."

42. Grant, *Raising Baby by the Book,* 15.

43. Ulrich, "Housewife and Gadder"; Osterud, "'She Helped Me Hay It as Good as a Man'"; Wermuth, "New York Farmers and the Market Revolution."

44. McMurray, *Transforming Rural Life.*

45. Golden, *A Social History of Wet Nursing in America;* Salmon, "The Cultural Significance of Breastfeeding and Infant Care."

46. Salmon, "The Cultural Significance of Breastfeeding and Infant Care," 21.

47. Salmon, "The Cultural Significance of Breastfeeding and Infant Care," 21.

48. Golden, *A Social History of Wet Nursing in America.*

49. Golden, *A Social History of Wet Nursing in America,* 43.

50. Fildes, *Breasts, Bottles and Babies.*

51. Mintz and Kellog, *Domestic Revolutions,* 55.

52. Riley, *Inventing the American Woman,* 68.

53. The history of the "republican mother" is widely discussed in the recent feminist scholarship on the state. See, for example, Baker, "The Domestication of Politics"; and Mink, "The Lady and the Tramp."

54. Riley, *Inventing the American Woman,* 46.

55. Riley, *Inventing the American Woman,* 46.

56. There is a large literature on the creation of the "private sphere." Some examples, besides those already cited, are Cott, *The Bonds of Womanhood;* Bloch, "American Feminine Ideals in Transition"; Ryan, *Cradle of the Middle Class;* and Baker, "The Domestication of Politics."

57. Mintz and Kellog, *Domestic Revolutions.*
58. Stansell, *City of Women.*
59. Mintz and Kellog, *Domestic Revolutions,* 63.
60. Heininger, *A Century of Childhood,* 21.
61. Heininger, *A Century of Childhood,* 21.
62. Heininger, *A Century of Childhood,* 19.
63. Heininger, *A Century of Childhood,* 20.
64. Heininger, *A Century of Childhood,* 21.
65. Clement, *Growing Pains: Children in the Industrial Age,* 63.
66. De Tocqueville, *Democracy in America.*
67. Martineau, in Kimmel and Stephen, *Social and Political Theory,* 117.
68. Cott, *The Bonds of Womanhood.*
69. Cott, *The Bonds of Womanhood.*
70. Mintz and Kellog, *Domestic Revolutions,* 63.
71. Smith-Rosenberg, "Beauty, the Beast, and the Militant Woman," 564.
72. Smith-Rosenberg, "Beauty, the Beast, and the Militant Woman."
73. Mintz and Kellog, *Domestic Revolutions,* 88.
74. Golden, *A Social History of Wet Nursing in America.*
75. Hareven, "Historical Changes in Children's Networks."
76. Clement, *Growing Pains,* 38.
77. Chezem and Friesen, "Attendance at Breast-Feeding Support Meetings," 83.
78. Quoted in Armstrong, "Choice Would Be a Fine Thing," 44.
79. Latour, *The Pasteurization of France,* 19.
80. Latour, *The Pasteurization of France,* 18.
81. Levenstein, *Revolution at the Table,* 135.

NOTES TO CHAPTER 4

1. Dillon, *Seven Decades of Milk,* 19.
2. Flexner, "The Battle for Pure Milk in New York City."
3. Cummings, *The American and His Food,* 90.
4. Cummings, *The American and His Food;* Rosenkrantz, "The Trouble with Bovine Tuberculosis" argues that the relationship between bovine and human tuberculosis was in fact minor.
5. Smith-Rosenberg, *Religion and the Rise of the American City,* 273, 272.
6. Danbom, *The Resisted Revolution,* 24.
7. Lloyd, "Lords of Industry," 548.
8. Danbom, *The Resisted Revolution,* 24.
9. Kirschner, *City and Country;* Barron, *Mixed Harvest.*
10. Danbom, *Born in the Country.*

11. Danbom, *The Resisted Revolution*, 36.

12. Saloutos and Hicks, *Twentieth Century Populism*, 22.

13. Cited in Danbom, *The Resisted Revolution*, 43; see also Flamm, "The National Farmers Union and the Evolution of Agrarian Liberalism" for an account of NFU farmers' political adjustment to the rising political power of consumers and Roosevelt's consumerist New Deal policies.

14. Spencer and Blanford, *An Economic History of Milk Marketing and Pricing*.

15. U.S. Public Health Service, *Milk and Its Relation to Public Health*.

16. Cummings, *The American and His Food*, 94.

17. Eager, "Morbidity and Mortality Statistics."

18. Meckel, *Save the Babies*.

19. Straus, *Disease in Milk*. Interestingly enough, Straus's milk stations were criticized for encouraging the early weaning of babies: Levenstein, *Revolution at the Table*, 131.

20. Brunger, "Dairying and Urban Development," 173.

21. Wiebe, *The Search for Order*.

22. Yet, as a sociologist looking at the rise of a food habit, I realize that using the histories of reform movements to create a social context for my story requires constant vigilance. As is true of most histories of social change, there are different ways to tell each reform movement story. For every statement characterizing one or the other reform movement as based in a particular set of political motivations, there is another statement saying exactly the opposite. Historians have characterized each movement as alternatively conservative or progressive, humanitarian or self-interested, harking back to the old society or heralding the new. For each era, therefore, my examination of the pertinent social movements must walk a line between the intellectual dishonesty of simply choosing historical viewpoints that most easily agree with my own interpretation, and a more thorough examination of the historical controversies around the interpretation of these movements. Too much attention to alternative interpretations would turn the book into a historiography of social movements rather than about a particular food habit.

23. Leavitt, *The Healthiest City*; Pegram, "Public Health and Progressive Dairying in Illinois."

24. Shover, *First Majority, Last Minority*.

25. Brunger, "Dairying and Urban Development."

26. Rosenau, *The Milk Question*, 2.

27. The mandatory pasteurization law exempted certified milk, which comprised only a small percentage of the milk market, as well as manufacturing milk, which was not supposed to be sold for drinking purposes.

28. New York Milk Committee, *Proceedings: Conference on Milk Problems*, 9.

29. New York Milk Committee, *Proceedings: Conference on Milk Problems*, 11. Interestingly, milk reformers in Chicago did make the link between low rates of

breast-feeding and high infant mortality. They campaigned vigorously to encourage mothers to breast-feed, with little success. This campaign did not include lobbying for maternal work relief. Wolf, "Don't Kill Your Baby."

30. Ebling et al., *Wisconsin Dairying*, 24.

31. Wasserman, "Henry L. Coit and the Certified Milk Movement," 367. Certified milk in the 1890s cost twelve cents a quart versus six cents a quart for uncertified milk.

32. Giblin, *Milk: The Fight for Purity*, 42.

33. Spencer and Blanford, *An Economic History of Milk Marketing and Pricing*.

34. Meckel, *Save the Babies*, 87.

35. Meigs quoted in Meckel, *Save the Babies*, 82.

36. Meckel, *Save the Babies*, 48; Apple, *Mothers and Medicine*.

37. Meckel, *Save the Babies*, 49.

38. Meckel, *Save the Babies*, 50.

39. Wasserman, "Henry L. Coit and the Certified Milk Movement."

40. Lederle to New York Milk Committee, *Proceedings: Conference on Milk Problems*, 97.

41. North to New York Milk Committee, *Proceedings: Conference on Milk Problems*, 178.

42. North to New York Milk Committee, *Proceedings: Conference on Milk Problems*, 173.

43. North to New York Milk Committee, *Proceedings: Conference on Milk Problems*, 173.

44. Spencer and Blanford, *An Economic History of Milk Marketing and Pricing*.

45. Audience comment to New York Milk Committee, *Proceedings: Conference on Milk Problems*, 92.

46. President, Western New York Milk Producers' Association to New York Milk Committee, *Proceedings: Conference on Milk Problems*, 93–94.

47. Rosenau, *The Milk Question*, 246.

48. Spencer and Blanford, *An Economic History of Milk Marketing and Pricing*, 67.

49. Mortenson, *An Economic Study of the Milwaukee Milk Market*, 49.

50. Clement and Warber, *The Market Milk Business of Detroit*.

51. Spencer and Blanford, *An Economic History of Milk Marketing and Pricing*, 417.

52. Manchester, *The Public Role in the Dairy Economy*.

53. Chandler, *The Visible Hand*.

54. Rosenau, *The Milk Question*, 242.

55. Rosenau, *The Milk Question*, 242

56. Rosenau, *The Milk Question*, 242.

57. Rosenau, *The Milk Question*, 242.

58. Rosenau, *The Milk Question*, 253.

59. Moldenhawer testimony to New York Milk Committee, *Proceedings: Conference on Milk Problems*, 17, italics in original.

60. Norton and Spencer, *A Preliminary Survey of Milk Marketing*, 5.

61. Leavitt, *The Healthiest City*.

62. In 1913 the Supreme Court decided in favor of the city in *Adams v. Milwaukee*, the landmark case that enabled cities to control milk production outside city limits: see Leavitt, *The Healthiest City*.

63. Pegram, "Public Health and Progressive Dairying," 37.

64. Leavitt, *The Healthiest City*, 258.

65. Block, "The Development of Regional Institutions in Agriculture"; Block and DuPuis, "Making the Country Work for the City."

66. Spencer and Blanford, *An Economic History of Milk Marketing and Pricing*, 781.

NOTES TO CHAPTER 5

1. The technology for printing illustrated advertising had been available since the sixteenth century. However, most editors considered large, illustrated advertisements unethical, since they enabled larger companies to advertise their products more prominently than smaller businesses. Newspapers restricted advertisements to two lines, and magazine advertisements were not much larger. Larger companies eventually found ways around these restrictions, one of the earliest being repetitions of a two-line ad for Ivory soap—Ivory: It Floats—on several pages.

2. Merchant, *The Death of Nature*, 9.

3. Merchant, *The Death of Nature*, 20.

4. Chandler, *The Visible Hand*. Chandler discusses Borden's in *Scale and Scope*.

5. Hall, "New Ethnicities."

6. Martin, "The End of the Body?"

7. Greene, *Darwin and the Modern World View*, 43.

8. Conser, *God and the Natural World*, 6.

9. Conser, *God and the Natural World*, 6.

10. Cited in de Beer, 71.

11. Cosslett, *Science and Religion in the Nineteenth Century*, 86; Greene, *Darwin and the Modern World View*, 42.

12. For a description of Americans' search for a new worldview that incorporated industrial dynamism, see Wiebe, *The Search for Order*.

13. Greene, *Darwin and the Modern World View*, 46.

14. Gonce, "The Social Gospel, Ely, and Commons's Initial Stage of Thought."

15. Saum, *The Popular Mood of Pre–Civil War America*, 3.
16. Lincoln's First Inaugural Address.
17. Lincoln's Second Inaugural Address.
18. For example, many of the early condensed milks did not contain enough fat or vitamin C for infant feeding, leading to an increase in diseases of undernourishment, such as rickets.
19. Presbrey, *The History and Development of Advertising*.
20. Meckel, *Save the Babies*, 151.
21. Merchant, *The Death of Nature*.
22. This was followed a few decades later by the American Dairy Association, which also carried out research and promotion.
23. Tauer and Forker, *Dairy Promotion in the United States*, 7.
24. Tauer and Forker, *Dairy Promotion in the United States*, 11–14.
25. Hoover and Hall, *Educational Milk-for-Health Campaigns*, 1.
26. Hoover and Hall, *Educational Milk-for-Health Campaigns*, 1.
27. Pirtle, *History of the Dairy Industry*, 133.
28. Hoover and Hall, *Educational Milk-for-Health Campaigns*, 2.
29. Chinn, *Health Habits*, 3.
30. Chinn, *Health Habits*, 3.
31. Hartford Public Schools Department of Food Services and Nutrition Education, Web page.
32. Chinn, *Health Habits*, 15.
33. Chinn, *Health Habits*, 13–15.
34. Steenbock and Hart, *Milk, the Best Food*.
35. National Dairy Council, *Milk Made the Difference*.
36. National Dairy Council, *Milk Made the Difference*, 4.
37. Petty, "Food, Poverty and Growth," 38.
38. Tauer and Forker, *Dairy Promotion in the United States*.
39. Levenstein, *Paradox of Plenty*, 59.
40. Levenstein, *Paradox of Plenty*, 59.
41. Levenstein, *Paradox of Plenty*, 59
42. Levenstein, *Paradox of Plenty*, 59.
43. Norton and Spencer, *A Preliminary Survey of Milk Marketing*, 5.
44. Cummings, *The American and His Food*.
45. Johnston, *A Hundred Years of Eating*.
46. Brody, "Debate over Milk: Time to Look at Facts."
47. McIntosh, *American Food Habits in Historical Perspective*.
48. Zuckerman, *The Potato*.
49. McIntosh, *American Food Habits in Historical Perspective*.
50. Becker, "Will Milk Make Them Grow?"
51. National Dairy Council, *Milk Made the Difference*. This quote occurs in numerous National Dairy Council publications from the 1920s.

52. Hedrick, *A History of Agriculture in the State of New York*, 362–63.

53. Pillsbury, *No Foreign Food*.

54. Warren, testimony in *Report of the Joint Legislative Committee to Investigate the Milk Industry*.

55. Mortenson, *Milk Distribution as a Public Utility*.

NOTES TO CHAPTER 6

1. Belcher, *Clean Milk*.

2. Park in Belcher, *Clean Milk*, 19.

3. Rabinbach, *The Human Motor*.

4. Locke, *Second Treatise*.

5. Danbom, *The Resisted Revolution*, 29.

6. Current survey results are published under Cornell's "Dairy Farm Business" series.

7. Stephens, *Economic Studies of Dairy Farming in New York*, part 11, 40.

8. Warren, *Factors for Success on Dairy and General Farms*, 22.

9. Warren, *Factors for Success on Dairy and General Farms*, 78. In years of low prices, the bulletins' authors had a harder time making this argument, since the data showed that larger farms lost their advantage during these periods.

10. Warren, *Factors for Success on Dairy and General Farms*, 78.

11. Stephens, *Economic Studies of Dairy Farming in New York*, part 11, 7.

12. Mann and Dickinson, "Obstacles to the Development of a Capitalist Agriculture."

13. Reducing seasonality was also important for reasons of milk pricing that will be discussed in chapter 8.

14. Cunningham, *Commercial Dairy Farming: Oneida-Mohawk Region*, 38.

15. Cunningham, *Commercial Dairy Farming: Oneida-Mohawk Region*, 38.

16. Misner, *Economic Studies of Dairy Farming in New York*, part 13, *100 Grade A Farms in the Tully-Homer Area*, 27.

17. Misner, *Economic Studies of Dairy Farming in New York*, part 13, *100 Grade A Farms in the Tully-Homer Area*, 27.

18. For example, Cunningham, *Seasonal Costs and Returns in Producing Milk*; Parsons, *Effect of Changes in Milk and Feed Prices*.

19. Cunningham, *Commercial Dairy Farming: Plateau Region*, 40. Cunningham's bulletins for the Hudson Valley and Oneida-Mohawk regions make similar recommendations.

20. Bierly, *Factors That Affect Costs and Returns in Producing Milk*, 20.

21. Pirtle, *History of the Dairy Industry*.

22. Pirtle, *History of the Dairy Industry*, 33.

23. Atack and Bateman, *To Their Own Soil*, 147.

24. Atack and Bateman, *To Their Own Soil*, 151.

25. Lampard, *The Rise of the Dairy Industry in Wisconsin.*

26. Schmidt and Ross, *Readings in the Economic History of American Agriculture*, 337.

27. Rogers and Shoemaker, *Communication of Innovations.*

28. Rogers and Shoemaker, *Communication of Innovations.* For a critique of this literature, see Geisler and DuPuis, "From Green Revolution to Gene Revolution."

29. For example, Cunningham, *Commercial Dairy Farming: Hudson Valley Region.*

30. Dairy Competitiveness Project Interviews, dairy farmer interviews. Thomas Lyson, director. Melanie DuPuis, Roger Stoll, and James Cruise, interviewers. Cattarangus and Allegany counties, 1987.

31. Beck, *Types of Farming in New York*, pg. 20–21.

32. Interestingly, while much of agricultural research has focused on increasing the productivity of plants on good soils, this was not the case with alfalfa research.

33. Warren, *Factors for Success on Dairy and General Farms*, 19.

34. Bierly, *Factors That Affect Costs and Returns in Producing Milk.*

35. Warren, testimony in *Report of the Joint Legislative Committee to Investigate the Milk Industry*, 39.

36. Stephens, *Economic Studies of Dairy Farming in New York*, part 11, *Success in Management of Dairy Farms as Affected by the Proportion of the Factors of Production.*

37. Cunningham, *Commercial Dairy Farming in New York.*

38. Cunningham, *Commercial Dairy Farming: Plateau Region*, 15.

39. Nash, *Wilderness and the American Mind.*

40. Geisler and Lyson, "The Cumulative Impact of Dairy Industry Restructuring."

41. Dairy Competitiveness Project Interviews, 1987.

42. Outlaw et al., *Structure of the U.S. Dairy Farm Sector.*

43. Kelleher and Bills, *Statistical Summary of the 1987 Farm Management and Energy Survey.*

44. Manchester, *The Public Role in the Dairy Economy.*

45. Wisconsin Statistical Reporting Service, *Wisconsin Dairy Herds.* These were Grade B herds.

46. Smith, *The Wealth of Nations*, 423.

NOTES TO CHAPTER 7

1. Friedmann, "The Family Farm and the International Food Regimes"; Mann and Dickinson, "Obstacles to the Development of a Capitalist Agriculture"; Buttel, "Beyond the Family Farm."

2. Bennet, *Of Time and the Enterprise*; Salamon, "Ethnic Communities and the Structure of Agriculture"; Salamon and O'Reilly, "Family Land and Developmental Cycles."

3. Cruise and Lyson, "Beyond the Farmgate." See also the work of the British rural geographers Terry Marsden, Richard Munton, and Sarah Whatmore.

4. See Simmonds, *The State of the Art of Farming Systems Research* for an overview of this approach.

5. Simmonds, *The State of the Art of Farming Systems Research.*

6. Altieri, *Agroecology: Science of Sustainable Agriculture*; Gliessman, *Agroecology.*

7. Granovetter, "Economic Action and Social Structure."

8. Gilbert and Akor, "Increased Structural Divergence in U.S. Dairying"; Lyson and Geisler, "Toward a Second Agricultural Divide."

9. Goodman, "Agro-Food Studies in the 'Age of Ecology.'"

10. Busch et al., "Beyond Political Economy"; Whatmore and Thorne, "Nourishing Networks."

11. Rossetto, preface to van der Ploeg and Long, *Born from Within.*

12. There were a number of other dairy farm strategies, including milk for butter production from mixed "general" farms. Although the analysis does contain some of these data, the comparison is primarily based on a comparison of cheese and market milk dairy areas, because these systems of dairying were historically the most distinct.

13. Salamon, *Prairie Patrimony.*

14. Another indicator would be acres of corn silage. Unfortunately, historical township-level data on this indicator were not available.

15. McIntosh, *American Food Habits in Historical Perspective.*

16. Catherwood, *A Statistical Study of Milk Production*, 24.

17. Catherwood, *A Statistical Study of Milk Production*, 25.

18. Misner, *Economic Studies of Dairy Farming in New York*, part 5, *Cheese-Factory Milk*, 39.

19. Misner, *Economic Studies of Dairy Farming in New York*, part 5, *Cheese-Factory Milk*, pg. 50.

20. Bierly, *Factors That Affect Costs and Returns in Producing Milk*, 3.

21. Bierly, *Factors That Affect Costs and Returns in Producing Milk*, 15.

22. McMurray, *Transforming Rural Life.*

23. Roadhouse and Henderson, *The Market-Milk Industry*; Pirtle, *History of the Dairy Industry.*

24. Chandler, *Scale and Scope.*

25. Bakken, *American Cheese Factories in Wisconsin*, 5.

26. Brunger, "Dairying and Urban Development."

27. Spencer and Blanford, *An Economic History of Milk Marketing and Pricing*, 25.

28. *Rural New Yorker,* August 3, 1861, quoted in Spencer and Blanford, *An Economic History of Milk Marketing and Pricing,* 26.

29. Spencer and Blanford, *An Economic History of Milk Marketing and Pricing,* 86.

30. Calculated from Manchester, *The Public Role in the Dairy Economy,* 11.

31. Spencer and Blanford, *An Economic History of Milk Marketing and Pricing,* 375.

32. The condensery system, being large-scale due to the capital investment needed to create the product, yet creating a product with a somewhat longer shelf life, was a hybrid of the two systems. Gail Borden's built the first condensed milk plant in 1856 in Wolcottville, Connecticut.

33. Gilbert and Akor, "Increased Structural Divergence in U.S. Dairying"; Lyson and Geisler, "Toward a Second Agricultural Divide"; Lyson and Gillespie, "Producing More Milk on Fewer Farms."

34. Outlaw et al., *Structure of the U.S. Dairy Farm Sector.*

35. Lyson and Gillespie, "Producing More Milk on Fewer Farms."

36. Lyson and Gillespie, "Producing More Milk on Fewer Farms." Because many farmers kept a cow for personal use early in the century, the number of farms reporting milk cows does not accurately represent the number of dairy farms. However, definitions of dairy farm varied greatly over this period and therefore pose difficulty in terms of representing changes over time.

37. Gilbert and Akor, "Increased Structural Divergence in U.S. Dairying."

38. Outlaw et al., *The Structure of the U.S. Dairy Farm Sector.*

39. Kalter et al., *Biotechnology and the Dairy Industry;* Kalter, "The New Biotech Agriculture: Unforeseen Economic Consequences"; Lyson and Gillespie, "Producing More Milk on Fewer Farms."

40. Benjamin, "Theses on the Philosophy of History."

41. Long and van der Ploeg, "Endogenous Development"; Cruise and Lyson, "Beyond the Farmgate"; Lockie and Kitto, "Beyond the Farm Gate."

42. Childs, "Transcommunality"; Hall, "New Ethnicities."

43. Van der Ploeg, "Styles of Farming."

44. Long and van der Ploeg, "Endogenous Development," 1.

45. Long and van der Ploeg, "Endogenous Development."

46. Van der Ploeg, "Styles of Farming," 18.

47. One example of policy analysis that typified this perspective was the highly influential report by the Office of Technology Assessment, *Technology, Public Policy, and the Changing Structure of American Agriculture.*

48. Rossetto, preface to vander Ploeg and Long, *Born from Within,* vii.

49. Rossetto, preface to vander Ploeg and Long, *Born from Within,* viii.

NOTES TO CHAPTER 8

1. The description of the 1939 strike comes from Kriger, "Syndicalism and Spilled Milk," and "The 1939 Dairy Farmers Union Strike."

2. This order was later expanded to include New Jersey and is now referred to as the New York–New Jersey milk market order.

3. Kriger, "The 1939 Dairy Farmers Union Strike."

4. Brunger, "Dairying and Urban Development," 171.

5. Brunger, "Dairying and Urban Development," 172.

6. Brunger, "Dairying and Urban Development," 172.

7. Friedmann, "The International Relations of Food"; Raynolds, "The Restructuring of Export Agriculture"; Friedland, "The Global Fresh Fruit and Vegetable System."

8. Porter, *The Competitive Advantage of Nations.*

9. Harrison, *Lean and Mean.*

10. For example, Global Exchange is an activist group attempting to create coalitions between workers at a global level through its Global Sweatshop campaign.

11. Schlebecker, in "The World Metropolis and the History of American Agriculture," critiques the common argument that farmer disunity was due to "excessive individualism." Instead, using von Thünen's ideas, he posits that the farmer's interest was significantly influenced by a farm's spatial location: "[t]he different interests of farmers in different zones may have been more decisive than any amount of individualism."

12. Bensel, *Sectionalism and American Political Development*; the two classic statements on American sectionalism are Turner, "Sections and Nation," and Key, *Politics, Parties, and Pressure Groups.*

13. Lorence, *Gerald J. Boileau and the Progressive-Farmer-Labor Alliance.*

14. Young, "Does the American Dairy Industry Fit the Meso-Corporatist Model?"; Odom, "Associated Milk Producers."

15. Dillon, *Seven Decades of Milk*; Abrahams, "Agricultural Adjustment during the New Deal"; Barron, *Mixed Harvest.*

16. Colman, "Theoretical Models and Oral History Interviews"; Spencer and Blanford, *An Economic History of Milk Marketing and Pricing.*

17. Spencer and Blanford, *An Economic History of Milk Marketing and Pricing.*

18. Spencer and Blanford, *An Economic History of Milk Marketing and Pricing*; Dillon, *Seven Decades of Milk*; Colman, "Theoretical Models and Oral History Interviews."

19. Spencer and Blanford, *An Economic History of Milk Marketing and Pricing.*

20. Kriger, "The 1939 Dairy Farmers Union Strike," 6.

21. Kriger, "The 1939 Dairy Farmers Union Strike."

22. Dillon, *Seven Decades of Milk*; Kriger, "The 1939 Dairy Farmers Union Strike"; Barron, *Mixed Harvest*.

23. *New York Times*, August 13, 1933, sec. 4, 1, cited in Kriger, "The 1939 Dairy Farmers Union Strike."

24. Kriger, "The 1939 Dairy Farmers Union Strike."

25. Kriger, "The 1939 Dairy Farmers Union Strike."

26. One exception to this is the Wisconsin milk strike of 1933. While the strike was centered in the state's eastern fluid milk production counties, manufacturing milk farmers also withheld their milk. Hoglund, "Wisconsin Dairy Farmers on Strike."

27. Spencer and Blanford, *An Economic History of Milk Marketing and Pricing*.

28. Spencer, *Cooperative Marketing of Milk in New York*; Pollard and Champlin, *Receipts of Milk and Cream at the New York Market*.

29. Spencer and Blanford, *An Economic History of Milk Marketing and Pricing*, 357.

30. Colman, "Theoretical Models and Oral History Interviews."

31. Bond, "Organization Problems of Marginal and Border-Line Territory," 2.

32. Bond, "Organization Problems of Marginal and Border-Line Territory," 1.

33. Black, *The Dairy Industry and the AAA*; Cassels, *A Study of Fluid Milk Prices*; von Thünen, *The Isolated State*. For a more detailed examination of the role of von Thünen in the establishment of market order legislation, see Block and DuPuis, "Making the Country Work for the City."

34. Von Thünen, *The Isolated State*.

35. Black, *The Dairy Industry and the AAA*, 154.

36. Harvey, "The Spatial Fix."

37. Manchester, *The Public Role in the Dairy Economy*.

38. Bond, "Organization Problems of Marginal and Border-Line Territory."

39. Manchester, *The Public Role in the Dairy Economy*.

40. McConnell, *The Decline of Agrarian Democracy*; Kriger, "The 1939 Dairy Farmers Union Strike."

41. Barron, *Mixed Harvest*.

42. Spencer and Blanford, *An Economic History of Milk Marketing and Pricing*.

43. Dillon, *Seven Decades of Milk*; Spencer and Blanford, *An Economic History of Milk Marketing and Pricing*.

44. Colman, "Theoretical Models and Oral History Interviews"; Spencer and Blanford, *An Economic History of Milk Marketing and Pricing*.

45. Dillon, *Seven Decades of Milk*.

46. Ford, "Another Double Burden," 385.

47. Ford, "Another Double Burden," 385.

48. Hoovers Online Company Profile.

49. Suiza Foods, 1999 Annual Report.
50. Federal Reserve Bank report; Craig Alexander, personal communication.
51. "Judiciary Panel Extends Northeast Dairy Compact."
52. "A Contented Moo," 28.
53. "A Contented Moo," 28.
54. The literature on "border studies" makes a similar point about culture. See, for example, Rosaldo, *Culture and Truth,* and Anzaldúa, *Borderlands/La Frontera.*

NOTES TO CHAPTER 9

1. Odom, "Associated Milk Producers."
2. Dillon, *Seven Decades of Milk.* In fact, the only truly successful New York state milk strike—in 1916, which succeeded in cutting the city's milk supply in half—occurred during a time when dealers were unable to find alternate sources of milk, due to the massive World War I exports of dairy products to Europe.
3. Manchester, *The Public Role in the Dairy Economy.*
4. Christopherson, "Market Rules and Territorial Outcomes."
5. See in particular Sanders, *Roots of Reform.*
6. Ritter, *Goldbugs and Greenbacks;* Sanders, *Roots of Reform.*
7. Abrahams, "Agricultural Adjustment During the New Deal"; Kriger, "The 1939 Dairy Farmers Union Strike"; Barron, *Mixed Harvest.*
8. Sanders, *Roots of Reform,* 16.
9. Sanders, *Roots of Reform.*
10. Stonecash, "Political Parties and Legislative Behavior in New York." Suburbanization has changed these alliances somewhat, especially on Long Island.
11. Bailey, *Marketing and Pricing of Milk and Dairy Products in the United States.* Bailey adds that these measures were not important, since production was local. However, these ordinances were imposed for a reason, in this case the setting of boundaries within the local milkshed. For example, according to a contemporary source, city inspection regulations were becoming "increasingly rigid, and it is an effective economic barrier against the bringing in of milk supplied from outside of the established sheds." Ebling et al, *Wisconsin Dairying,* 44. With the increased movement of milk in the 1950s and 1960s, most of these ordinances were challenged in court and declared unconstitutional. See *Dean Milk Company v. City of Madison* 340 U.S. 349, 1951.
12. *Farmland Dairies et al. v. Commissioner of New York State Department of Agriculture and Markets* 847 F. 2d 1038, 1988.
13. Grossman, "Milk Fight."

14. Colman, *Education and Agriculture.*

15. Spencer and Blanford, *An Economic History of Milk Marketing and Pricing.* Colman, "Theoretical Models and Oral History Interviews"; Gould Colman, personal communication, 1987.

16. The Agricultural Adjustment Act of 1933 first instituted milk marketing orders, but they were not fully authorized until the Agricultural Marketing Agreement Act of 1937.

17. Williams et al., *Organization and Competition in the Midwest Dairy Industry,* 14.

18. Decades of court challenges to aspects of market orders that could be considered a constraint of interstate trade (and therefore unconstitutional) eventually made market orders less capable of forming restrictive boundaries. As a result, over the years, market order policies became more similar from one order to the next. Recent federal legislation formally standardized in law a number of the more distinctive policies described above.

19. Jacobs, "Debates in Rural Land Planning Policy."

20. Jacobs, "Debates in Rural Land Planning Policy."

21. Ladd, "Land Planning in the Empire State," 306.

22. Jacobs, "Debates in Rural Land Planning Policy."

23. National Resources Planning Board, *A Report on National Planning and Public Works,* part 2, 181

24. National Resources Planning Board, *A Report on National Planning and Public Works,* part 2, 182.

25. National Resources Planning Board, *A Report on National Planning and Public Works,* part 2, 183.

26. Gee, "Rural Population Research in Relation to Land Utilization," 357.

27. Gee, "Rural Population Research in Relation to Land Utilization," 357.

28. Lane, *Submarginal Farm Lands in New York State,* 13.

29. Lane, *Submarginal Farm Lands in New York State,* iii.

30. Lane, *Submarginal Farm Lands in New York State,* iii.

31. Wooten, *The Land Utilization Program,* 1.

32. Park, "Human Migration and the Marginal Man."

33. DuPuis, "In the Name of Nature," and "The Land of Milk,"; Jacobs, "Debates in Rural Land Planning Policy"; Colman, *Education and Agriculture.* A portion of Tompkins County did eventually become a Resettlement Administration project.

34. Hill, "Land Classification Tour."

35. Guttenberg, "The Land Utilization Movement of the 1920's," 482.

36. Cited in Colman, *Education and Agriculture,* 276.

37. Burrit Papers, box 2414.

38. Burrit Papers, box 2414.

39. For example, on February 13, 1939, Burrit wrote to Farm Light and Power Committee general secretary Edward S. Foster, "I recognize that it is the support of organized agriculture that has had a great deal to do both with my original appointment and with the re appointment." Burrit Papers, Cornell University, box 2414.

40. Burrit Papers, box 2414.

41. Burrit Papers, box 2414.

42. Lane, *Submarginal Farm Lands in New York State*, 4.

43. See, for example, Abdalla et al., "What We Know about Historical Trends"; Lanyon, "Implications of Dairy Herd Size."

44. Sanders, *Roots of Reform.*

45. Saloutos and Hicks, *Twentieth Century Populism*, 34.

46. See McConnell, *The Decline of Agrarian Democracy* for a history of the Farm Bureau's rise to prominence.

47. Hoglund, "Wisconsin Dairy Farmers on Strike."

48. Hoglund, "Wisconsin Dairy Farmers on Strike."

49. Kolb and Day, *Interdependence in Town and Country Relations*, 38.

50. Many cheese factories had started out as farmers' cooperatives in New York state, but later these plants were bought by local merchants.

51. Urban fluid milk markets have been so small in the state of Wisconsin that figures on this usage have historically been included under the title "Other Uses" in state dairy reports.

52. Taylor, *The History of Agricultural Economics in the United States*, part 5.

53. Hoglund, "Wisconsin Dairy Farmers on Strike."

54. McIntyre, *Fifty Years of Cooperative Extension in Wisconsin*, 82.

55. McIntyre, *Fifty Years of Cooperative Extension in Wisconsin*, 68.

56. McIntyre, *Fifty Years of Cooperative Extension in Wisconsin*, 72.

57. Saloutos and Hicks, *Twentieth Century Populism.*

58. Saloutos and Hicks, *Twentieth Century Populism*; McIntyre, *Fifty Years of Cooperative Extension in Wisconsin*; Glover, *Farm and College*, 332–33; Taylor and Taylor, *The Story of Agricultural Economics in the United States.*

59. Hoglund, "Wisconsin Dairy Farmers on Strike."

60. Saloutos and Hicks, *Twentieth Century Populism.*

61. McIntyre, *Fifty Years of Cooperative Extension in Wisconsin*, 82; Chambers, *California Farm Organizations.*

62. Schaars interview, 36.

63. See *Dean Milk Company v. City of Madison*, 1951.

64. Williams et al., *Organization and Competition in the Midwest Dairy Industry.*

65. Williams et al., *Organization and Competition in the Midwest Dairy Industry*, 66.

66. Williams et al., *Organization and Competition in the Midwest Dairy Industry*, 80.

67. Williams et al., *Organization and Competition in the Midwest Dairy Industry*, 74.

68. Ebling et al., *Wisconsin Dairying*, 37; Williams et al., *Organization and Competition in the Midwest Dairy Industry*.

69. Andrew Novakovic, personal communication, 2000.

70. Parks, "Experiment in the Democratic Planning of Public Agricultural Activity."

71. University of Wisconsin, *Outline for County Land Use Planning in Wisconsin*, 1; italics in original.

72. Guttenberg, "The Land Utilization Movement of the 1920's," 483.

73. Even these farms do not fit the industrial ideal of pristine barns and white uniforms: the largest tend to be smelly feed lots with a milking parlor attached.

74. For an overview, see Butler, *Do State/Local Regulations Interfere*.

75. Butler, *Do State/Local Regulations Interfere*.

76. Liebman, *California Farmland*; Chambers, *California Farm Organizations*.

77. Fielding, "The Los Angeles Milkshed."

78. Jeffrey, *The Production-Consumption Balance of Milk in the Northeast Dairy Region*; Luke, *Utilization and Pricing of Milk under the New York Milk Marketing Order*.

79. Fielding, "The Los Angeles Milkshed"; Fletcher and McCorkle, *Growth and Adjustment of the Los Angeles Milkshed*.

80. Copley, "A Historical Geography of the Dairy Industry of Stanislaus County, California," 146.

81. Fielding, "The Los Angeles Milkshed."

82. Fielding, "The Los Angeles Milkshed," 8.

83. Copley, "A Historical Geography of the Dairy Industry of Stanislaus County, California," 44.

84. Kuhrt, *The Story of California's Milk Stabilization Laws*; Butler, *Do State/Local Regulations Interfere*.

85. California agricultural producers avoid referring to themselves as farmers.

86. However, the California order also sets higher fluid milk prices for pasture-based dairy systems in the northern coastal region.

87. While retail price setting was eliminated in the 1970s, California still uses its unique regional cost of production models to set prices at the producer level.

88. Fielding, "The Los Angeles Milkshed."

89. Access to fluid milk markets has extended far beyond these allotments in recent years, although this system is still in place.

90. Leon Garoyan, personal communication, 1990.

91. Friedland's studies of lettuce and tomatoes show how agricultural institutions in California primarily played a reactive, service role to the state's

agricultural interests. See Friedland et al., *Manufacturing Green Gold,* and *Destalking the Wily Tomato.*

92. Garoyan, personal communication, 1990.
93. Howard Conklin, personal communication, 1990.
94. Conklin, personal communication, 1990.
95. Weeks et al., *Land Utilization in the Northern Sierra Nevada,* 113.
96. Weeks et al., *Land Utilization in the Northern Sierra Nevada,* 114.
97. Weeks et al., *Land Utilization in the Northern Sierra Nevada,* 75.
98. See, for example, Outlaw et al., *Structure of the U.S. Dairy Farm Sector;* Kalter et al., *Biotechnology and the Dairy Industry;* Office of Technology Assessment, *Technology, Public Policy, and the Changing Structure of American Agriculture.*
99. As the controversy developed, scientists began to refer to this substance as BST, or "bovine somatotropin," a term that avoids the hormone stigma.
100. See, for example, Barbano, "What's the Fuss about Cow Hormones?"
101. Kalter et al., *Biotechnology and the Dairy Industry,* 82.
102. Kalter et al., *Biotechnology and the Dairy Industry,* 112.
103. Buttel and Geisler, "The Recombinant BGH Controversy."
104. Buttel and Geisler, "The Social Impacts of Bovine Somatotropin."
105. Butler, *Do State/Local Regulations Interfere.* California agricultural officials moved from cost of production pricing, based mostly on feed costs, to formula pricing in response to a growing rejection of the cost-based pricing system by farmers themselves, particularly the Western United Dairymen Association. Bees Butler, personal communication.
106. Bees Butler, interview by author, October 16, 2000.
107. Liebhardt, *The Dairy Debate.*
108. Buttel, "The Recombinant BGH Controversy"; and Buttel and Geisler, "The Social Impacts of Bovine Somatotropin."
109. Many of the officials who implemented the New York dairy policies described in this chapter eventually went on to international posts. Cornell's School of Agriculture became the training ground for people interested in World Bank or USAID careers.
110. Another anti-agricultural policy commonly implemented has been the purchase of crops dumped on the market at low prices by developed countries, which undercuts local producers' prices for these foods. See Lappé and Collins, *Food First.*

NOTES TO CHAPTER 10

1. Jacobsen and Outlaw, "Dairy Product Consumption and Demand," 1.
2. Buttel, "The Recombinant BGH Controversy," 6.

3. Barboza, "Modified Foods Put Companies in a Quandary"; Montague,"Biotech in Trouble—Part 2."

4. For a more contemporary example, see McMenamin and McNamara, *Milking the Public.*

5. France, "Groups Debate the Role of Milk in Building a Better Pyramid."

6. France, "Groups Debate the Role of Milk."

7. USDA, *Nutrition and Your Health: Dietary Guidelines for Americans.*

8. The Web site is www.purefood.org.

9. Hightower, "Food Monopoly," 3.

10. See, for example, Anderson, *Genetic Engineering, Food and Our Environment.*

11. Buttel, "The Recombinant RBGH Controversy"; Buck et al., "From Farm to Table"; Guthman, "Regulating Meaning, Appropriating Nature."

12. Cohen, *Milk: The Deadly Poison.*

13. Cohen, "Dairy Education Board Founder Blames Flo Jo's Death on Dairy Products."

14. Bourdieu's ideas about "habitus" have been helpful to my thinking here. Bourdieu, *Distinction.*

15. Gramsci, *Selections from the Prison Notebooks.*

16. Ordonez, "Cash Cows: Hamburger Joints Call Them 'Heavy Users.'"

17. For example, the early USDA film, "Milk for You and Me." National Agricultural Library.

18. Scott, "Organic Dairy a Cash Cow."

19. Wyngate, "Organic Dairy: The Little Niche That Could"; Scott, "Organic Dairy a Cash Cow."

20. Murphy, "More Buyers Are Asking: Got Milk without Chemicals?"

21. Gilbert, "Fears over Milk."

22. Hoover, "The Other Milky Way."

23. Belasco, *Appetite for Change.*

24. Scott, "Organic Dairy a Cash Cow."

25. Friedland, "The New Globalization"; Buck et al., "From Farm to Table"; Guthman, "Regulating Meaning, Appropriating Nature"; Marsden and Arce, "Constructing Quality"; Kenney et al., "Agriculture in U.S. Fordism."

26. Belasco, *Appetite for Change*; Whatmore and Thorne, "Nourishing Networks."

27. Belasco, *Appetite for Change*; Whatmore and Thorne, "Nourishing Networks."

28. Bourdieu, *Distinction*; Goldfrank, "Fresh Demand: The Consumption of Chilean Produce in the United States."

29. Buck et al., "From Farm to Table," 16. I've heard organic farmers refer to salad mix as "yuppie chow."

30. Goodman and Redclift, *Refashioning Nature.*

31. Mintz, *Sweetness and Power.*

32. Friedland, "The New Globalization"; Marsden and Arce, "Constructing Quality."

33. Marsden and Wrigley, "Regulation, Retailing and Consumption," 1911.

34. Kauffman, "New Age Meets New Right."

35. Buck et al., "From Farm to Table."

36. In contrast, the top fifty general food processing firms in the United States represent 47 percent of sales. U.S. Department of Commerce, Bureau of Economic Analysis, *Survey of Current Business.*

37. Looker, "High-Level Turbulence."

38. *Dairy Products News,* May 3, 1999.

39. Manchester and Blayney, *The Structure of Dairy Markets,* 14–15.

40. Manchester and Blayney, *The Structure of Dairy Markets,* 36.

41. Manchester and Blayney, *The Structure of Dairy Markets,* 41.

42. "Organic Opportunities."

43. See Szasz, *Inverted Quarantine* for an analysis of risk politics in terms of the elite's search for an "inverted quarantine."

44. See Miller, *Acknowledging Consumption,* and Featherstone, *Consumer Culture and Postmodernism* for overviews. A number of food system analysts have called for more incorporation of a cultural perspective on consumption into food system studies. See Goodman, "Agro-Food Studies in the 'Age of Ecology'"; Marsden and Arce, "Constructing Quality"; Fine et al., *Consumption in the Age of Affluence;* Buttel, "The Recombinant BGH Controversy."

45. Goodman, "Agro-Food Studies in the 'Age of Ecology,'" 32

46. Weber, "Power, Domination, and Legitimacy."

47. Woods, "Rethinking Elites."

48. The idea of politics as a struggle over representation is taken from Stuart Hall and Chantal Mouffe.

49. Callon, "Some Elements of a Sociology of Translation"; Latour, "The Powers of Association." In relation to food system studies, Goodman has called for a rethinking of the framework in terms of this network approach. Goodman, "Agro-Food Studies in the 'Age of Ecology.'"

50. Ulrich Beck, Anthony Giddens, and Christopher Lash have developed the idea of "reflexive modernization." My characterization of reflexive consumption is similar. However, their ideas appear to be caught up in a fight over the usefulness of postmodernism to sociology and a reassertion of ideas of modernity. I have no intellectual investment in that argument, one way or another. The idea of reflexivity simply helps me understand the politics of consumption. Lash and Urry have also developed a notion of reflexive consumption that emphasizes urban taste and style as forms of knowledge but

understates the extent to which reflexivity and discursive activities still relate to traditional political issues such as agrarianism or environmentalism. Lash and Urry, *Economies of Signs and Space.*

51. Allen and Sachs, "Sustainable Agriculture in the United States."

52. Brown, "Popular Epidemiology."

53. Wiles et al., "A Shopper's Guide to Pesticides in Produce."

54. The Center for Food Safety has formally petitioned the FDA to remove rBGH from the market because it allegedly ignored the results of particular animal health studies. Health Canada also charged the FDA with ignoring these studies. For more information, see CFS News online at http://www.centerforfoodsafety.org/factsandissues.

55. Newman, "CFC Phase-Out Moving Quickly"; Hinrichsen, "Stratospheric Maintenance."

56. Stipp, "Is Monsanto's Biotech Worth Less Than a Hill of Beans?"

57. It would have been interesting to see whether a non-European country could have had the same impact on Monsanto's stock price. Racism and the struggle for legitimacy are not necessarily separate issues. The reflexive consumer may not necessarily be a middle-class white person with a college education and European descent, but the enrollment discourse below will show that dominant actors often target this group. The loss of this group's support in many industry sectors spells disaster for the industry. As a result, the sustainable food discourse has generally ignored issues of race: Allen and Sachs, "Sustainable Agriculture in the United States."

58. See the articles in Krimsky and Golding, *Social Theories of Risk*; National Research Council, *Understanding Risk*; Beck, *Risk Society*; Perrow, *Normal Accidents.*

59. Slovic, "Public Perception of Risk," and "Perceptions of Risk: Reflections on the Psychometric Paradigm"; Perrow, *Normal Accidents.*

60. "1997 Grand Makeover."

61. "1997 Grand Makeover."

62. "Can We Have a Safe, Secure Food Supply?"

63. *Rural Vermont Report*, July–August 1999, 6.

64. Cummins, "Organic Standards: Who Really Speaks for the Organic Consumer?"

Bibliography

Abdalla, Charles W., Les E. Lanyon, and Milton C. Hallberg. "What We Know about Historical Trends in Firm Location Decisions and Regional Shifts: Policy Issues for an Industrializing Animal Sector." *American Journal of Agricultural Economics* 77 (5) (December 1995): 1229–37.

Abrahams, Paul. "Agricultural Adjustment during the New Deal Period: The New York Milk Industry: A Case Study." *Agricultural History* 39 (2) (1965): 92–101.

Ahlstrom, Sydney E. *A Religious History of the American People*. New Haven: Yale University Press, 1972.

Allen, Ann Taylor. *Feminism and Motherhood in Germany, 1800–1914*. New Brunswick: Rutgers University Press, 1991.

Allen, Patricia, and Carolyn Sachs. "Sustainable Agriculture in the United States: Engagements, Silences, and Possibilities for Transformation." In *Food for the Future: Conditions and Contradictions of Sustainability*, ed. Patricia Allen. New York: John Wiley, 1993.

Altieri, Miguel. *Agroecology: Science of Sustainable Agriculture*. Boulder: Westview, 1995.

Alvord, Henry. "Dairy Development in the United States." In *Yearbook of the United States Department of Agriculture, 1899*, 381–98. Washington, DC: GPO, 1900.

Anderson, Luke. *Genetic Engineering, Food and Our Environment*. White River Junction, VT: Chelsea Green, 1999.

Anderson, Oscar. *The Health of a Nation: Harvey J. Wiley and the Fight for Pure Food*. Chicago: University of Chicago Press, 1958.

Anzaldúa, Gloria. *Borderlands/La Frontera*. 2d ed. San Francisco: Aunt Lute, 1999.

Apple, Rima D. *Mothers and Medicine: A Social History of Infant Feeding, 1890–1950*. Madison: University of Wisconsin Press, 1987.

Armstrong, Sue. "Choice Would Be a Fine Thing: An Increasingly Urban Lifestyle Is Encouraging Women to Abandon Breast-Feeding." *New Scientist* 48 (1998): 44–46.

Atack, Jeremy, and Fred Bateman. *To Their Own Soil: Agriculture in the Antebellum North*. Ames: Iowa State University Press, 1987.

Atkins, P. J. "White Poison? The Social Consequences of Milk Consumption, 1850–1930." *Social History of Medicine* 5 (2) (1992): 207–27.

Bailey, Kenneth W. *Marketing and Pricing of Milk and Dairy Products in the United States*. Ames: Iowa State University Press, 1997.

Baker, Burton A., and Rudolph K. Froker. *The Evaporated Milk Industry under Federal Marketing Agreements*. Bulletin 156. Madison: University of Wisconsin Agricultural Experiment Station, 1940.

Baker, Paula. *The Moral Frameworks of Public Life: Gender, Politics, and the State in Rural New York, 1870–1930*. New York: Oxford University Press, 1991.

———. "The Domestication of Politics: Women and American Political Society, 1780–1920." In *Women, the State, and Welfare*, ed. Linda Gordon. Madison: University of Wisconsin Press, 1990.

Bakken, Henry H. *American Cheese Factories in Wisconsin*. Bulletin 156. Madison: University of Wisconsin Agricultural Experiment Station, 1930.

Ball, Helen H., and Alan Swedlund. "Poor Women and Bad Mothers: Placing the Blame for Turn-of-the-Century Infant Mortality." *Northeast Anthropology* 52 (1996): 31–52.

Barbano, David M. "What's the Fuss about Cow Hormones?" *Consumers' Research Magazine* 77 (5) (1994): 14–18.

Barboza, David. "Modified Foods Put Companies in a Quandary." *New York Times*, June 3, 2000.

Barkun, Michael. *Crucible of the Millennium: The Burned-Over District of New York in the 1840s*. Syracuse: Syracuse University Press, 1986.

Barron, Hal. *Mixed Harvest: The Second Great Transformation in the Rural North, 1870–1930*. Chapel Hill: University of North Carolina Press, 1997.

Barwell, Mrs. (Louisa Mary). *Advice to Mothers on the Treatment of Infants with Directions for Self-Management before, during and after Pregnancy*. Philadelphia: Leary and Getz, 1853.

Baumslag, Naomi, and Dia L. Michels. *Milk, Money and Madness: The Culture and Politics of Breastfeeding*. Westport, CT: Bergin and Garvey, 1995.

Beck, Martin L. *Types of Farming in New York*. Bulletin 704. Ithaca: Cornell University Agricultural Experiment Station, 1962.

Beck, Ulrich. *Risk Society: Toward a New Modernity*, trans. Mark Ritter. Thousand Oaks, CA: Sage, 1992.

Becker, Stanley. "Will Milk Make Them Grow? An Episode in the Discovery of the Vitamins." *ACS Symposium Series* 228 (1983): 61–83.

Belasco, Warren J. *Appetite for Change: How the Counterculture Took on the Food Industry*. Ithaca: Cornell University Press, 1993.

Belcher, Sarah Drowne. *Clean Milk*. With an introduction by William H. Park. New York: Hardy Publishing Company, 1903.

Benjamin, Walter. "Theses on the Philosophy of History." In *Illuminations*. New York: Schocken, 1969.

Bennet, John W. *Of Time and the Enterprise: North American Family Farm Management in the Context of Resource Marginality*. Minneapolis: University of Minnesota Press, 1982.

Bensel, Richard Franklin. *Sectionalism and American Political Development: 1880–1980*. Madison: University of Wisconsin Press, 1984.

Bidwell, Percy Wells. "The Agricultural Revolution in New England." *American Historical Review* 26 (July 1921): 683–702.

Bidwell, Percy Wells, and John I. Falconer. *History of Agriculture in the Northern United States, 1620–1860*. New York: P. Smith, 1941.

Bierly, Ivan R. *Factors That Affect Costs and Returns in Producing Milk*. Bulletin 804. Ithaca: Cornell University Agricultural Experiment Station, 1944.

Bird, Elizabeth Ann. "The Social Construction of Nature: Theoretical Approaches to the History of Environmental Problems." *Environmental Review* 11 (4) (1987): 255–64.

Black, John D. *The Dairy Industry and the AAA*. Washington, DC: Brookings Institution, 1935.

Bloch, Ruth H. "American Feminine Ideals in Transition: The Rise of the Moral Mother, 1785–1815." *Feminist Studies* 4 (1978): 101–26.

Block, Daniel Ralston. "The Development of Regional Institutions in Agriculture: The Chicago Milk Marketing Order." Ph. D. diss, University of California at Los Angeles, 1997.

Block, Daniel Ralston, and E. Melanie DuPuis. "Making the Country Work for the City." *American Journal of Economics and Sociology*, forthcoming.

Bogart, Ernest L. *Economic History of American Agriculture*. Wilmington, DE: Scholarly Resources, 1973 [1923].

Bond, M. C. "Organization Problems of Marginal and Border-Line Territory in Reference to Metropolitan Milk Sheds." In *Economic Extension Manual*, part 4, *Dairy Cooperatives and Credit*. Ithaca: Cornell University Department of Agricultural Economics, 1932–33. Mimeo.

Bosselman, Fred, and David Callies. *The Quiet Revolution in Land Use Control*. Washington, DC: Council of Environmental Quality, 1971.

Bourdieu, Pierre. *Distinction: A Social Critique of the Judgement of Taste*. Cambridge: Harvard University Press, 1984.

Brinton, Mary C., and Victor Nee, eds. *The New Institutionalism in Sociology*. New York: Russell Sage Foundation, 1998.

Brody, Jane. "Debate over Milk: Time to Look at Facts." *New York Times* September 26, 2000, D8.

Brown, Phil. "Popular Epidemiology Challenges the System." *Environment* 35, (8) (October 1993): 16–31.

Brumberg, Joan Jacobs. "The Appetite as Voice." In *Food and Culture,* ed. Carole Counihan and Penny Van Esterick. New York: Routledge, 1997.

Brunger, Eric. "Dairying and Urban Development in New York State, 1850–1900." *Agricultural History* 29 (4)(1955): 169–73.

Bryant, Carol A., et al. *The Cultural Feast: An Introduction to Food and Society.* St. Paul: West Publishing Company, 1985.

Buck, D., C. Getz, and J. Guthman. "From Farm to Table: The Organic Vegetable Commodity Chain of Northern California." *Sociologia Ruralis* 37 (1)(1997): 3, 16.

Buikstra, J. E., L. W. Konigsberg, and J. Bullington. "Fertility and the Development of Agriculture in the Prehistoric Midwest." *American Antiquity* 51 (1986): 528–46.

Burrit, Maurice Chase. Papers. Cornell University Archives, Ithaca, NY.

Busch, L., K. Tanaka, and A. Juska. "Beyond Political Economy: Actor Networks and the Globalization of Agriculture." *Review of International Political Economy* 4 (4) (1997): 688–708.

Butler, L. J. *Do State/Local Regulations Interfere with the Federal Milk (Price Support) Program? A Case Study: California Pricing.* Washington, DC: National Commission on Dairy Policy, 1988.

Buttel, Frederick H. "The Recombinant BGH Controversy in the United States: Toward a New Consumption Politics of Food?" *Agriculture and Human Values* 17 (3) (2000): 5–20.

———. "Beyond the Family Farm." In *Technology and Social Change in Rural Areas,* ed. G. F. Summers. Boulder: Westview, 1983.

Buttel, Frederick, and Charles Geisler. "The Social Impacts of Bovine Somatotropin: Emerging Issues." In *Biotechnology and the New Agricultural Revolution,* ed. Joseph J. Molnar and Henry Kinnucan. Boulder: Westview, 1989.

Callon, M. "Some Elements of a Sociology of Translation: Domestication of the Scallops and the Fishermen of St. Breux Bay." In *Power, Action and Belief: A New Sociology of Knowledge?* Ed. J. Law. London: Routledge and Kegan Paul, 1986.

Campbell, Colin. *The Romantic Ethic and the Spirit of Consumerism.* New York: Blackwell, 1987.

"Can We Have a Safe, Secure Food Supply with Just a Handful of Huge Food Corporations in Control?" *Regional Farm and Food Project* 2 (2) (1999): 1.

Carper, Steve. *Milk Is Not For Every Body: Living with Lactose Intolerance.* New York: Plume, 1996.

Cartensen, Vernon. *Farms or Forests: Evolution of a State Land Policy for Northern Wisconsin, 1850–1932.* Madison: University of Wisconsin, 1958.

Cassels, John M. *A Study of Fluid Milk Prices.* Cambridge: Harvard University Press, 1937.

Catherwood, M. P. *A Statistical Study of Milk Production for the New York Market.* Bulletin 518. Ithaca: Cornell University Agricultural Experiment Station, 1931.

Chambers, Clarke A. *California Farm Organizations: A Historical Study of the Grange, the Farm Bureau, and the Associated Farmers, 1929–1941.* Berkeley: University of California Press, 1952.

Chandler, Alfred Dupont. *Scale and Scope: The Dynamics of Industrial Capitalism.* Cambridge, MA: Belknap, 1990.

———. *The Visible Hand: The Managerial Revolution in American Business.* Cambridge, MA: Belknap, 1977.

Chayanov, A. V. *A Theory of Peasant Economy.* Madison: University of Wisconsin Press, 1990.

Chezem, Jocarol, and Carol Friesen. "Attendance at Breast-Feeding Support Meetings: Relationship to Demographic Characteristics and Duration of Lactation in Women Planning Postpartum Employment." *Journal of the American Dietetic Association* 99 (1) (1999): 83–86.

Childs, John Brown. "Transcommunality: From the Politics of Conversion to the Ethics of Respect in the Context of Cultural Diversity—Learning from Native American Philosophies with a Focus on the Haudensosaunee." *Social Justice* 25 (1998) : 143–69.

Chinn, Aubyn. *Health Habits: Suggestions for Developing Them in School Children.* Chicago: National Dairy Council, 1924.

Christopherson, Susan. "Market Rules and Territorial Outcomes: The Case of the United States." *International Journal of Urban and Regional Research* 17 (2) (1993): 274–89.

Clarke, Adele, and Theresa Montini. "The Many Faces of RU486: Tales of Situated Knowledges and Technological Contestations." *Science, Technology, and Human Values* 18 (1) (1993): 42–79.

Clawson, Marion. *New Deal Planning: The National Resources Planning Board.* Baltimore: Johns Hopkins University Press, 1981.

———. *Forests for Whom and for What?* Baltimore: Johns Hopkins University Press, 1975.

Clement, C. E., and G. P. Warber. *The Market Milk Business of Detroit, Michigan, in 1915.* Bulletin 639. Washington, DC: USDA, 1918.

Clement, Priscilla Ferguson. *Growing Pains: Children in the Industrial Age, 1850–1890.* New York: Twayne, 1997.

Cochrane, Willard Wesley. *The Development of American Agriculture: A Historical Analysis.* Minneapolis: University of Minnesota Press, 1979.

Cohen, Robert. "Dairy Education Board Founder Blames Flo Jo's Death on Dairy Products." http://www.notmilk.com/deb/pr4.html. (November 13, 1998).

———. *Milk: The Deadly Poison.* Englewood Cliffs, NJ: Argus, 1997.

Collet, David. "Pastoralists and Wildlife: Image and Reality in Kenya Maasailand." In *Conservation in Africa: People, Policies, and Practice,* ed. David Anderson and Richard Grove. Cambridge: Cambridge University Press, 1987.

Colman, Gould. "Theoretical Models and Oral History Interviews." *Agricultural History* 41 (3) (1967): 255–66.

———. *Education and Agriculture: A History of the New York State College of Agriculture at Cornell University*. Ithaca: Cornell University Press, 1963.

Conser, Walter H., Jr. *God and the Natural World: Religion and Science in Antebellum America*. Columbia: University of South Carolina Press, 1993.

"A Contented Moo." *Economist*, August 8, 1998, 28–29.

Copley, Richard. "A Historical Geography of the Dairy Industry of Stanislaus County, California." Master's Thesis, Department of Geography, University of California at Berkeley, 1961.

Cosslett, Tess, ed. *Science and Religion in the Nineteenth Century*. New York: Cambridge University Press, 1984.

Cott, Nancy F. *The Bonds of Womanhood: "Woman's Sphere" in New England, 1780–1835*. New Haven: Yale University Press, 1997.

———. *The Grounding of Modern Feminism*. New Haven: Yale University Press, 1987.

Cross, Whitney. *The Burned-Over District: The Social and Intellectual History of Enthusiastic Religion in Western New York, 1800–1850*. New York: Farrar, Straus and Giroux, 1981.

Cruise, James, and Thomas Lyson. "Beyond the Farmgate: Factors Related to Agricultural Performance in Two Dairy Communities." *Rural Sociology* 56 (1) (1991): 41–55.

Cummings, Richard Osborn. *The American and His Food: A History of Food Habits in the United States*. Rev. ed. Chicago: University of Chicago Press, 1941.

Cummins, Ronnie. "Organic Standards: Who Really Speaks for the Organic Consumer?" *Food Bytes: News and Analysis on Genetic Engineering and Factory Farming*, April 8, 1998. www.purefood.org/Organic/foodByt8.htm.

Cunningham, L. C. *Commercial Dairy Farming: Hudson Valley Region, New York, 1961–62*. Bulletin 999. Ithaca: Cornell University Agricultural Experiment Station, 1964.

———. *Commercial Dairy Farming: Oneida-Mohawk Region, New York, 1959–60*. Bulletin 992. Ithaca: Cornell University Agricultural Experiment Station, 1964.

———. *Commercial Dairy Farming: Plateau Region, New York, 1957–58*. Bulletin 966. Ithaca: Cornell University Agricultural Experiment Station, 1961.

———. *Commerical Dairy Farming in New York*. Bulletin 857. Ithaca: Cornell University Agricultural Experiment Station, 1949.

———. *Seasonal Costs and Returns in Producing Milk in Orange County, New York*. Bulletin 641. Ithaca: Cornell University Agricultural Experiment Station, 1936.

Danbom, David. *The Resisted Revolution: Urban America and the Industrialization of Agriculture*. Ames: Iowa State University Press, 1979.

Danbom, David B. *Born in the Country: A History of Rural America.* Baltimore: Johns Hopkins University Press, 1995.

Davis, David Brion. *Ante-Bellum Reform.* New York: Harper and Row, 1967.

de Beer, Gavin. "Introduction." *Autobiographies/ Charles Darwin, Thomas Henry Huxley.* New York: Oxford University Press, 1983.

DeJanvry, Alain. *The Agrarian Question and Reformism in Latin America.* Baltimore: Johns Hopkins University Press, 1981.

Demos, John. "The Changing Faces of Fatherhood." In *Past, Present and Personal: The Family and the Life Course in American History.* New York: Oxford University Press, 1986.

De Tocqueville, Alexis. *Democracy in America.* New York: Bantam, 2000.

Dillon, John J. *Seven Decades of Milk: A History of New York's Dairy Industry.* New York: Orange Judd, 1941.

Douglas, Ann. *The Feminization of American Culture.* New York: Knopf, 1977.

Duffy, John. *A History of Public Health in New York City, 1866–1966.* New York: Russell Sage Foundation, 1974.

———. *A History of Public Health in New York City, 1625–1866.* New York: Russell Sage Foundation, 1968.

DuPuis, E. Melanie. "In the Name of Nature: Ecology, Marginality, and Rural Land Use Planning during the New Deal." In *Creating the Countryside: The Politics of Rural and Environmental Discourse,* ed. E. Melanie DuPuis and Peter Vandergeest. Philadelphia: Temple University Press, 1995.

———. "No Limits to Growth?" *California Farmer* 263 (September 7, 1995): 6, 17.

———. "Sub-National State Institutions and the Organization of Agricultural Resource Use: The Case of the Dairy Industry." *Rural Sociology* 58 (3) (1993): 440–60.

———. "The Land of Milk: Economic Organization and the Politics of Space in the U.S. Dairy Industry." Ph.D. diss., Department of Rural Sociology, Cornell University, 1991.

DuPuis, E. Melanie, and Charles Geisler. "Biotechnology and the Small Farm." *BioScience* 38 (6) (1988): 406–11.

Eager, J. M. "Morbidity and Mortality Statistics as Influenced by Milk." In *Milk and Its Relation to Public Health.* National Institutes of Health. Treasury Department, Public Health and Marine-Hospital Service Hygienic Laboratory. Bulletin 41. Washington, DC: GPO, 1908.

Ebling, Walter H., W. D. Bormuth, and F. J. Graham. *Wisconsin Dairying.* Statistical Bulletin 200. Madison: Wisconsin Crop Reporting Service, 1938.

Eckles, Clarence Henry, Willes Barnes Combs, and Harold Macy. *Milk and Milk Products.* 4th ed. New York: McGraw-Hill, 1957.

Eckles, C. H., and G. F. Warren. *Dairy Farming.* New York: Macmillan, 1916.

Ellis, David M., James A. Frost, Harold C. Syrett, and Harry J. Carman. *A History of New York State.* Ithaca: Cornell University Press, 1957.

Erdman, H. E. "The 'Associated Dairies' of New York as Precursors of American Agricultural Cooperatives." *Agricultural History* 36 (2) (1962): 82–90.

Featherstone, Mike. *Consumer Culture and Postmodernism.* London: Sage, 1991.

Fenton, Alexander. "Milk Products in the Everyday Diet of Scotland." In *Milk and Milk Products: From Medieval to Modern Times*, ed. Patricia Lysaght. Edinburgh: Canongate Academic, 1994.

Fielding, Gordon. "The Los Angeles Milkshed: A Study of the Political Factor in Agriculture." *Geographical Review* 54 (1964): 1–12.

———. "Dairying in Cities Designed to Keep People Out." *Professional Geographer* 14 (1) (1962): 13–17.

Fildes, Valerie. "The Culture and Biology of Breastfeeding: An Historical Review of Western Europe." In *Breastfeeding: Biocultural Perspectives*, ed. Patricia Stuart-Macadam and Katherine A. Dettwyler. New York: Aldine de Gruyter, 1995.

———. "Breast-Feeding in London, 1905–19." *Journal of Biosocial Science* 24 (1)(1992): 53–70.

———. *Breasts, Bottles and Babies: A History of Infant Feeding.* Edinburgh: Edinburgh University Press, 1986.

Fine, Ben, Michael Heasman, and Judith Wright. *Consumption in the Age of Affluence: The World of Food.* New York: Routledge, 1996.

Flamm, Michael W. "The National Farmers Union and the Evolution of Agrarian Liberalism, 1937–1946." *Agricultural History* 68 (3): 54–81.

Fletcher, Lehman B., and Chester O. McCorkle, Jr. *Growth and Adjustment of the Los Angeles Milkshed.* Bulletin 787. Berkeley: California Agricultural Experiment Station, 1962.

Flexner, James. "The Battle for Pure Milk in New York City." In New York City Department of Health Milk Commission, *Is Loose Milk a Hazard?* Appendix. New York: Department of Health, 1931.

Ford, Linda G. "Another Double Burden: Farm Women and Agrarian Activism in Depression Era New York State." *New York History* 75 (4) (1994): 373–96.

Foucault, Michel. *The History of Sexuality: An Introduction*, vol. 1. New York: Vintage, 1978.

France, David. "Groups Debate the Role of Milk in Building a Better Pyramid." *New York Times*, June 29, 1999.

Friedland, William H. "The Global Fresh Fruit and Vegetable System." In *The Global Restructuring of Agro-Food Systems*, ed. Philip McMichael. Ithaca: Cornell University Press, 1994.

———. "The New Globalization: The Case of Fresh Produce." In *From Columbus to ConAgra*, ed. Alejandro Bonnano. Kansas City: University Press of Kansas, 1994.

Friedland, Willaim H., Amy E. Barton, and Robert J. Thomas. *Manufacturing Green Gold: Capital, Labor and Technology in the Lettuce Industry.* New York: Cambridge University Press, 1978.

Friedland, William H. and Amy Barton, with the assistance of Robert J. Thomas and Vicki Bolam. *Destalking the Wily Tomato: A Case Study in Social Consequences in California Agricultural Research.* Research monograph. Department of Applied Behavioral Science 15, University of California at Davis, 1975.

Friedmann, Harriet. "The International Relations of Food." In *Food: Multidisciplinary Perspectives,* B. Harris-White and R. Hoffenberg. Oxford: Blackwell, 1994.

———. "The Political Economy of Food." *New Left Review* 197 (1993): 29–57.

———. "The Family Farm and the International Food Regimes." In *Peasant and Peasant Societies,* ed. T. Shanin. Oxford: Blackwell, 1987.

———. *Are Distributions Really Structures?: A Critique of the Methodology of Max Weber.* Research Paper 63. Toronto: Centre for Urban and Community Studies, University of Toronto, 1974.

Gates, Paul W. "An Overview of American Land Policy." *Agricultural History* 50(1) (1976): 213–29.

———. *The Farmer's Age: Agriculture, 1815–1860.* New York: Holt, Rinehart, and Winston, 1960.

Gee, Wilson. "Rural Population Research in Relation to Land Utilization." *Social Forces* 12 (3) (1934): 355–59.

Geisler, Charles, and Melanie DuPuis. "From Green Revolution to Gene Revolution: What We Can Learn from New Biotechnology Strategies in the Third World." In *Biotechnology and the New Agricultural Revolution,* ed. Joseph J. Molnar and Henry Kinnucan. Boulder: Westview, 1989.

Geisler, Charles, and Thomas Lyson. "The Cumulative Impact of Dairy Industry Restructuring." *BioScience* 41 (8)(1991): 560–68.

Giblin, James. *Milk: The Fight for Purity.* New York: Crowell, 1986.

Gibson, A., and T. C. Smout. "From Meat to Meal: Changes in Diet in Scotland." In *Food, Diet and Economic Change Past and Present,* ed. Catherine Geissler and Derek J. Oddy. Leicester, UK: Leicester University Press, 1993.

Gilbert, Benjamin Davis. *The Cheese Industry of the State of New York.* Washington, DC: USDA, Bureau of Animal Industry, 1896.

Gilbert, Jess. "Democratic Planning in Agricultural Policy: The Federal-County Land-Use Planning Program, 1938–1942." *Agricultural History* 70 (2) (1996): 233–51.

Gilbert, Jess, and Raymond Akor. "Increased Structural Divergence in U.S. Dairying: California and Wisconsin since 1950." *Rural Sociology* 53 (1988): 56–72.

Gilbert, Susan. "Fears over Milk, Long Dismissed, Still Simmer." *New York Times*, January 19, 1999, F7.

Gliessman, Stephen R., ed., *Agroecology: Researching the Ecological Basis for Sustainable Agriculture*. New York: Springer-Verlag, 1989.

Glover, W. H. *Farm and College: The College of Agriculture at the University of Wisconsin: A History*. Madison: University of Wisconsin Press, 1952.

Golden, Janet. *A Social History of Wet Nursing in America: From Breast to Bottle*. New York: Cambridge University Press, 1996.

Goldfrank, Walter L. "Fresh Demand: The Consumption of Chilean Produce in the United States." In *Commodity Chains and Global Capitalism*, ed. Gary Gereffi and Miguel Korzeniewicz. Westport, CT: Praeger, 1994.

Gonce, R. A. "The Social Gospel, Ely, and Commons's Initial Stage of Thought." *Journal of Economic Issues* 30 (3) (1996): 641–66.

Goodman, David. "Agro-Food Studies in the 'Age of Ecology': Nature, Corporeality, Bio-politics." *Sociologia Ruralis* 39 (1) (1999): 17–38.

Goodman, David, and Michael Redclift. *Refashioning Nature: Food, Ecology, and Culture*. London: Routledge, 1991.

Goodman, David, Bernardo Sorj, and John Wilkinson. *From Farming to Biotechnology: A Theory of Agro-Industrial Development*. New York: Blackwell, 1987.

Goody, Jack. *Cooking, Cuisine, and Class: A Study in Comparative Sociology*. New York: Cambridge University Press, 1982.

Gramsci, Antonio. *Selections from the Prison Notebooks*, ed. and trans. Geoffrey Nowell Smith and Quintin Hoare. New York: International, 1971.

Granovetter, Mark. "Economic Institutions as Social Constructions: A Framework for Analysis." *Acta Sociologica* 35 (1) (1992): 3–12.

———. "Economic Action and Social Structure: The Problem of Embeddedness." *American Journal of Sociology* 91 (3) (1985): 481–510.

Grant, Julia. *Raising Baby by the Book: The Education of American Mothers*. New Haven: Yale University Press, 1998.

Greene, John C. *Darwin and the Modern World View*. New York: New American Library, 1963.

Griscom, John H. *The Sanitary Condition of the Laboring Population of New York*. New York: Harper and Bros., 1845.

Grossman, John. "Milk Fight: When Farmland Dairies Broke Up the New York City Milk Cartel, the Real War Began." *Inc.*, September 1987, 37–41.

Gusfield, Joseph R. *Symbolic Crusade: Status Politics and the American Temperance Movement*. Urbana: University of Illinois Press, 1963.

Guthman, J. "Regulating Meaning, Appropriating Nature: The Codification of California Organic Agriculture." *Antipode* 30 (2) (1998): 135+.

Guttenberg, Albert A. "The Land Utilization Movement of the 1920's." *Agricultural History* 50 (1976): 477–90.

Hall, Stuart. "New Ethnicities." In *Stuart Hall: Critical Diologues in Cultural Studies*, ed. David Morley and Kuan-Hsing Chen. New York: Routledge, 1996.

Haraway, Donna J. *Primate Visions: Gender, Race and Nature in the World of Modern Science*. New York: Routledge, 1989.

Hareven, Tamara. "Historical Changes in Children's Networks in the Family and the Community." In *Children's Social Networks and Social Supports*, ed. Deborah Belle. New York: Wiley, 1989.

Harrison, Bennett. *Lean and Mean: The Changing Landscape of Corporate Power in the Age of Flexibility*. New York: Basic Books, 1994.

Hartford Public Schools Department of Food Services and Nutrition Education. *Nutrition News* 6 (6), 1998–1999. Web page. www.nai.net/~foodserv/NEWS599.htm.

Hartley, Isaac Smithson. *Memorial of Robert Milham Hartley*. New York: Arno Press, 1976.

Hartley, Robert Milham. *An Historical, Scientific and Practical Essay on Milk as an Article of Human Sustenance*. New York: Arno Press, 1977 [1842].

Harvey, David. "The Spatial Fix: Hegel, von Thünen, and Marx." *Antipode* 13 (3) (1981): 1–12.

Hawkins, Ann, and Frederick Buttel. "The Political Economy of 'Sustainable Development.'" Presentation at the Annual Meeting of the American Sociological Association, San Francisco, August 1989.

Hays, Samuel P. *Conservation and the Gospel of Efficiency: The Progressive Conservation Movement, 1890–1920*. Cambridge: Harvard University Press, 1959.

Hedrick, Ulysses Prentis. *A History of Agriculture in the State of New York*. Albany: New York State Agriculture Society, 1933.

Heininger, Mary Lynn Stevens. *A Century of Childhood, 1820–1920*. Rochester, NY: Margaret Woodbury Strong Museum, 1984.

Hightower, Jim. "Food Monopoly." In *The Big Business Reader: On Corporate America*, ed. Mark Green, Michael Waldman, and Robert K. Massie, Jr. New York: Pilgrim Press, 1983.

Hill, F. F. "Land Classification Tour." Ithaca: Cornell University Department of Agricultural Economics and Farm Management, 1932. Mimeo.

Hinrichs, Clare. "Consuming Images: Making and Marketing Vermont as a Distinctive Rural Place." In *Creating the Countryside: The Politics of Rural and Environmental Discourse*, ed. E. Melanie DuPuis and Peter Vandergeest. Philadelphia: Temple University Press, 1995.

Hinrichsen, Don. "Stratospheric Maintenance: Fixing the Ozone Hole Is a Work in Progress." *Amicus Journal* 18 (3) 1996: 35–39.

Hoffert, Sylvia. *Private Matters: American Attitudes toward Childrearing and Infant Nurture in the Urban North, 1800–1860*. Urbana: University of Illinois Press, 1989.

Hoglund, William. "Wisconsin Dairy Farmers on Strike." *Agricultural History* 35 (1961): 24–34.

Hoover, Barbara. "The Other Milky Way." *Detroit News* online, April 2, 1996. http://detnews.com/menu/food0402.htm.

Hoover, Jessie M, and Florence L. Hall. *Educational Milk-For-Health Campaigns*. USDA Department Circular 250. Washington, DC: USDA, 1927.

"Horizon Purchases Organic Cow Brand." *Food Ingredients Online*, April 27, 1999.

Jacobs, Harvey. "Debates in Rural Land Planning Policy: A Twentieth Century History from New York State." *Journal of Rural Studies* 5 (2) (1989): 137–48.

Jacobsen, Robert E., and Joe L. Outlaw. *Dairy Product Consumption and Demand*. Dairy Markets and Policy: Issues and Options Series, no. M-3. Ithaca: Cornell University Program on Dairy Markets and Policy, 1995.

Jeffrey, Arthur D. *The Production-Consumption Balance of Milk in the Northeast Dairy Region*. Bulletin A.E. 1055. Ithaca: Cornell University Department of Agricultural Economics, Cornell University Agricultural Experiment Station, 1957.

Jensen, Joan M. *Loosening the Bonds: Mid-Atlantic Farm Women, 1750–1850*. New Haven: Yale University Press, 1986.

Johnson, J. D., N. Kretchmer, and F. J. Simoons. "Lactose Malabsorption: Its Biology and History." *Advances in Pediatrics* 21 (1974): 197–237.

Johnston, James P. *A Hundred Years Eating: Food, Drink and the Daily Diet in Britain since the Late Nineteenth Century*. Montreal: Gill and Macmillan; McGill-Queens University Press, 1977.

"Judiciary Subpanel Extends Northeast Dairy Compact." *CongressDaily/A.M.*, July 30, 1999.

Kalter, R. J. "The New Biotech Agriculture: Unforeseen Economic Consequences." *Issues in Science and Technology* 2 (1985): 125–33.

Kalter, R. J., R. Milligan, W. Lesser, W. McGrath, and D. Bauman. *Biotechnology and the Dairy Industry: Production Costs and Commercial Potential of the Bovine Growth Hormone*. A.E. 85–20. Ithaca: Cornell University Department of Agricultural Economics, 1985.

Kauffman, L. A. "New Age Meets New Right: Tofu Politics in Berkeley." *Nation*, September 16, 1991, 294–97.

Kautsky, Karl. "The Agrarian Question." In *The Rural Sociology of the Advanced Societies*, ed. Frederick Buttel and Howard Newby. Montclair, NJ: Allanheld Osmun, 1980.

Kelleher, M. J., and N. L. Bills. *Statistical Summary of the 1987 Farm Management and Energy Survey*. A.E. 89–3. Ithaca: Cornell University Department of Agricultural Economics, 1989.

Kenney, Martin, Linda Lobao, Jim Curry, and R. Goe. "Agriculture in U.S. Fordism: The Integration of the Productive Consumer." In *Towards a*

New Political Economy of Agriculture, ed. William H. Friedland, Lawrence Busch, Frederick H. Buttel, and Alan P. Rudy. Boulder: Westview, 1991.

Key, V. O. *Politics, Parties, and Pressure Groups*. New York: Crowell, 1964.

Kintner, Hallie J. "Trends and Regional Differences in Breast Feeding in Germany from 1871 to 1937." *Journal of Family History* 10(2)(1985): 163–82.

Kimmel, Michael S., and Charles Stephen. *Social and Political Theory: Classical Readings*. Boston: Allyn and Bacon, 1998.

Kirschner, Don. *City and Country: Rural Responses to Urbanization in the 1920s*. Westport, CT: Greenwood, 1970.

Kisbán, Eszter. "Milky Ways on Milk-Days: The Use of Milk and Milk Products in Hungarian Foodways." In *Milk and Milk Products: From Medieval to Modern Times*, ed. Patricia Lysaght. Edinburgh: Canongate Academic, 1994.

Kloppenberg, Jack, Jr. "Social Theory and the De/Reconstruction of Agricultural Science: Local Knowledge for an Alternative Agriculture." *Rural Sociology* 56 (4) (1991): 519–48.

Kolb, John H., and LeRoy J. Day. *Interdependence in Town and Country Relations in Rural Society: A Study of Trends in Walworth County, Wisconsin*. Bulletin 172. Madison: University of Wisconsin, 1950.

Kriger, Thomas J. "The 1939 Dairy Farmers Union Strike in Heuvelton and Canton, New York: The Story in Words and Pictures." *Journal for Multi-Media History* 1 (1) (1998). http://www.albany.edu/jmmh/vol1no1/dairy1.

———. "Syndicalism and Spilled Milk: The Origins of Dairy Farmer Activism in New York State, 1936–1941." *Labor History* 38 (2–3) (1997): 266–86.

Krimsky, Sheldon, and Dominic Golding, eds. *Social Theories of Risk*. Westport, CT: Praeger, 1992.

Kuhrt, William J. *The Story of California's Milk Stabilization Laws: From Chaos to Stability in the California Milk Industry*. Bulletin 54:4. Sacramento: California State Department of Agriculture, 1965.

Ladd, Carl E. "Land Planning in the Empire State." *New Republic*, August 3, 1932, 306–7.

Laird, Pamela Walker. *Advertising Progress: American Business and the Rise of Consumer Marketing*. Baltimore: Johns Hopkins University Press, 1998.

LaMont, T. E. *Agricultural Production in New York, 1866 to 1937*. Bulletin 693. Ithaca: Cornell University Agricultural Experiment Station, 1938.

Lampard, Eric E. *The Rise of the Dairy Industry in Wisconsin: A Study in Agricultural Change, 1820–1920*. Madison: State Historical Society of Wisconsin, 1963.

Lane, Charles. *Submarginal Farm Lands in New York State: A Report to the New York State Planning Board*. Albany: New York State Planning Board, 1935.

Lanyon, L. E. "Implications of Dairy Herd Size for Farm Material Transport, Plant Nutrient Management, and Water Quality." *Journal of Dairy Science* 74 (1992): 334–44.

Lappé, Frances Moore, and Joseph Collins, with Cary Fowler. *Food First: Beyond the Myth of Scarcity.* Boston: Houghton Mifflin, 1977.

Lash, Scott, and John Urry. *Economies of Signs and Space.* London: Sage, 1994.

Latour, Bruno. *The Pasteurization of France.* Cambridge: Harvard University Press, 1988.

———. "The Powers of Association." In *Power, Action and Belief: A New Sociology of Knowledge?* ed. John Law. London: Routledge and Kegan Paul, 1986.

Leavitt, Judith Walzer. *Brought to Bed: Childbearing in America, 1750 to 1950.* New York: Oxford University Press, 1986.

———. *The Healthiest City: Milwaukee and the Politics of Health Reform.* Princeton: Princeton University Press, 1982.

Lenin, Vladimir Ilyich. *The Development of Capitalism in Russia: The Process of the Formation of a Home Market for Large-Scale Industry.* Moscow: Foreign Languages Publishing House, 1956.

Levenstein, Harvey A. *Paradox of Plenty: A Social History of Eating in Modern America.* New York: Oxford University Press, 1993.

———. *Revolution at the Table: The Transformation of the American Diet.* New York: Oxford University Press, 1988.

Lewis, A. B. *An Economic Study of Land Utilization in Tompkins County, New York.* A.E. 590. Ithaca: Cornell University Department of Agricultural Economics, 1934.

Liebhardt, William. *The Dairy Debate: Consequences of Bovine Growth Hormone and Rotational Grazing Technologies.* Davis: University of California, Sustainable Ag., Res., and Ed. Prog. 1993.

Liebman, Ellen. *California Farmland: A History of Large Agricultural Land Holdings.* Totowa, NJ: Rowman and Allanheld, 1983.

Liestøl, Knut, Margit Rosenberg, and Lars Walløe. "Breast-Feeding Practice in Norway, 1860–1984." *Journal of Biosocial Science* 20 (1988): 45–58.

Lincoln, Abraham. *Abraham Lincoln: A Documentary Portrait through his Speeches and Writings*, ed. Don E. Fehrenbacher. New York: New American Library, 1964.

Lloyd, Henry Demarest. "Lords of Industry." *North American Review* 331 (June, 1884): 535–54.

Locke, John. The *Second Treatise of Government.* New York: Macmillan, 1986.

———. *Some Thoughts Concerning Education.* London: Heinemann, 1964 [1693].

Lockie, Stewart, and Simon Kitto. "Beyond the Farm Gate: Production-Consumption Networks and Agri-Food Research." *Sociologia Ruralis* 40 (1) (2000): 3–19.

Long, Ann, and Jan Douwe van der Ploeg. "Endogenous Development: Practices and Perspectives." In *Born from Within: Practice and Perspectives of Endogenous Rural Development*, ed. Jan Douwe van der Ploeg and Ann Long. Assen, Netherlands: Van Gorcum, 1994.

Looker, Dan. "High-Level Turbulence: Big Dairies Regroup, but Some Milk Meisters Continue to Grow." *Successful Farming Online*, November 1998.

Lorence, James J. *Gerald J. Boileau and the Progressive-Farmer-Labor Alliance: Politics of the New Deal*. Columbia: University of Missouri Press, 1994.

Luke, H. Alan. *Utilization and Pricing of Milk under the New York Milk Marketing Order*. Bulletin 866. Ithaca: Cornell University Agricultural Experience Station, 1950.

Lysaght, Patricia, ed. *Milk and Milk Products: From Medieval to Modern Times*. Edinburgh: Canongate Academic, 1994.

Lyson, Thomas A., and Charles C. Geisler. "Toward a Second Agricultural Divide: The Restructuring of American Agriculture." *Sociologia Ruralis* 32 (1992): 248–63.

Lyson, Thomas A., and Gilbert W. Gillespie. "Producing More Milk on Fewer Farms: Neoclassical and Neostructural Explanations of Changes in Dairy Farming." *Rural Sociology* 60 (3) (1995): 493–504.

Manchester, Alden Coe. *The Public Role in the Dairy Economy: Why and How Governments Intervene in the Milk Business*. Boulder: Westview, 1983.

Manchester, Alden, and Don Blayney. *The Structure of Dairy Markets: Past, Present, Future*. Washington, DC: USDA, Economic Research Service, 1997.

Mann, S., and J. Dickinson. "Obstacles to the Development of a Capitalist Agriculture." *Journal of Peasant Studies* 5 (July 1978): 466–81.

Marsden, Terry and A. Arce. "Constructing Quality: Emerging Food Networks in the Rural Transition." *Environment and Planning A* 27 (1995): 1261–79.

Marsden, Terry, Jonathon Murdoch, Philip Lowe, Richard Munton, and Andrew Flynn. *Constructing the Countryside*. Boulder: Westview, 1991.

Marsden, Terry and N. Wrigley. "Regulation, Retailing and Consumption." *Environment and Planning A* 27 (1995): 1899–1912.

Martin, Emily. "The End of the Body?" *American Ethnologist* 19 (1) (1992): 121–41.

———. *The Woman in the Body: A Cultural Analysis of Reproduction*. Boston: Beacon, 1987.

Marx, Leo. *The Machine in the Garden: Technology and the Pastoral Ideal in America*. New York: Oxford University Press, 1964.

McClellan, Steve. "Theorizing New Deal Farm Policy: Broad Constraints of Capital Accumulation and the Creation of a Hegemonic Relation." In *Towards a New Political Economy of Agriculture*, ed. William H. Friedland, Lawrence Busch, Frederick H. Buttel, and Alan P. Rudy. Boulder: Westview, 1991.

McConnell, Grant. *The Decline of Agrarian Democracy.* Berkeley: University of California Press, 1953.

McIntosh, Elaine N. *American Food Habits in Historical Perspective.* Westport, CT: Praeger, 1995.

McIntyre, E. R. *Fifty Years of Cooperative Extension in Wisconsin, 1912–1962.* Madison: University of Wisconsin, 1962.

McMahon, Sarah Francis. "A Comfortable Subsistence: A History of Diet in New England, 1630–1850." Ph.D. diss., Brandeis University, 1982.

McMenamin, Michael, and Walter McNamara. *Milking the Public: Political Scandals of the Dairy Lobby from L.B.J. to Jimmy Carter.* Chicago: Nelson-Hall, 1980.

McMichael, Philip, ed. *The Global Restructuring of Agro-Food Systems.* Ithaca: Cornell University Press, 1994.

McMurry, Sally. *Transforming Rural Life: Dairying Families and Agricultural Change, 1820–1885.* Baltimore: Johns Hopkins University Press, 1995.

Meckel, Richard A. *Save the Babies: American Public Health Reform and the Prevention of Infant Mortality, 1850–1929.* Baltimore: Johns Hopkins University Press, 1990.

Meine, Carl. *Aldo Leopold: His Life and Work.* Madison: University of Wisconsin Press, 1988.

Melucci, Alberto. *Nomads of the Present: Social Movements and Individual Needs in Contemporary Society.* Philadelphia: Temple University Press, 1989.

Merchant, Carolyn. *Ecological Revolutions: Nature, Gender, and Science in New England.* Chapel Hill: University of North Carolina Press, 1989.

———. *The Death of Nature: Women, Ecology, and the Scientific Revolution.* San Francisco: Harper and Row, 1980.

Miller, Daniel. *Acknowledging Consumption: A Review of New Studies.* New York: Routledge, 1995.

Mink, Gwendolyn. "The Lady and the Tramp: Gender, Race and the Origins of the American Welfare State." In *Women, the State, and Welfare,* ed. Linda Gordon. Madison: University of Wisconsin Press, 1990.

Mintz, Sidney. *Sweetness and Power: The Place of Sugar in Modern History.* New York: Penguin, 1985.

Mintz, Steven, and Susan Kellog. *Domestic Revolutions: A Social History of American Family Life.* New York: Free Press, 1988.

Misner, E. G. *Economic Studies of Dairy Farming in New York,* part 13, *100 Grade A Farms in the Tully-Homer Area, Crop Year 1936.* Bulletin 696. Ithaca, NY: Cornell University Agricultural Experiment Station, 1937.

———. *Economic Studies of Dairy Farming in New York,* part 5, *Cheese-Factory Milk.* Bulletin 442. Ithaca: Cornell University Agricultural Experiment Station, 1925.

Montague, Peter. "Biotech in Trouble—Part 2." *Rachel's Environmental Health Weekly* 696 (May 11, 2000).

Moring, Beatrice. "Motherhood, Milk and Money: Infant Mortality in Pre-Industrial Finland." *Social History of Medicine* 11 (2) (1998): 177–96.

Mortenson, W. P. *Milk Distribution as a Public Utility*. Chicago: University of Chicago Press, 1940.

———. *An Economic Study of the Milwaukee Milk Market*. Research Bulletin 113. Madison: University of Wisconsin Agricultural Experiment Station, 1932.

Motarjemi, Y., F. Kafterstein, G. Moy, and F. Quevedo. "Contaminated Weaning Food: A Major Risk Factor for Diarrhea and Associated Malnutrition." *Bulletin. World Health Organization, 71* (1993): 79–92.

Mullaly, John. *The Milk Trade of New York and Vicinity, Giving an Account of the Sale of Pure and Adulterated Milk*. New York: Fowlers and Wells, 1853.

Murphy, Kate. "More Buyers Are Asking: Got Milk without Chemicals?" *New York Times*, August 1, 1999, 6.

Nash, Roderick. *Wilderness and the American Mind*. 3d ed. New Haven: Yale University Press, 1982.

National Commission on Food Marketing. *Organization and Competition in the Dairy Industry*. Technical Study no. 3. Washington, DC: GPO, 1966.

National Dairy Council. *Milk Made the Difference*. Chicago, [1920?].

———. *What Milk Will Do for Your Child*. Chicago, 1921.

———. *Food Facts: It Is the Sacred Right of Your Child to Be as Healthy as Knowledge Can Make Them*. Chicago, 1919.

National Institutes of Health. *Milk and Its Relation to the Public Health*. Treasury Department, Public Health and Marine-Hospital Service Bulletin 41. Washington, DC: GPO, 1908.

National Research Council. *Understanding Risk: Informing Decisions in a Democratic Society*. Washington, DC: National Academy Press, 1996.

National Resources Planning Board. *A Report on National Planning and Public Works in Relation to Natural Resources and Including Land Use and Water Resources with Finding and Recommendations*. Washington, DC: GPO, 1934.

———. *A Report on National Planning and Public Works in Relation to Natural Resources and Including Land Use and Water Resources with Finding and Recommendations*. Part 2, *Report of the Land Planning Committee*. Washington, DC: GPO, 1934.

Nelson, Marie C., and J. Rogers, eds. *Urbanisation and the Epidemiologic Transition*. Uppsala: Uppsala University, 1989.

Newman, Alan. "CFC Phase-Out Moving Quickly." *Environmental Science and Technology* 28 (1) (1994): 35–38.

Newman, J. "How Breast Milk Protects Newborns." *Scientific American* 273 (1995): 76–79.

Newton, Isaac. *Opticks*. 1730.

New York Board of Health. *Majority and Minority Reports of the Select Committee of the Board of Health*. New York: C. W. Baker, 1858.

New York Milk Committee. *Ten Years of Work, 1907–1916. Report of the New York Milk Committee*. New York, 1916.

———. *Proceedings: Conference on Milk Problems*. New York, 1910.

New York State Department of Agriculture. *Butter and Cheese Factories, Milk Stations and Condenseries in the State of New York*. Bulletin 8. Albany, NY, 1906.

"1997 Grand Makeover." *Publish RGB*, June 1997, www.publish.com/features/9706/makeover.

Nixon, Edgar B., ed. *Franklin D. Roosevelt and Conservation, 1911–1945*. Hyde Park, NY: General Services Administration, National Archives and Records Service, 1957.

Norton, L. J., and Leland Spencer. *A Preliminary Survey of Milk Marketing in New York*. Bulletin 445. Ithaca: Cornell University Agricultural Experiment Station, 1925.

Odom, E. Dale. "Associated Milk Producers, Incorporated: Testing the Limits of Capper–Volstead." *Agricultural History* 59 (1) (1985): 40–55.

Office of Technology Assessment. *Technology, Public Policy, and the Changing Structure of American Agriculture*. Washington, DC: U.S. Congress, 1986.

Okun, Mitchell. *Fair Play in the Marketplace: The First Battle for Pure Food and Drugs*. Dekalb: Northern Illinois University Press, 1986.

Ordonez, Jennifer. "Cash Cows: Hamburger Joints Call Them 'Heavy Users.'" *Wall Street Journal*, January 12, 2000, A1.

"Organic Opportunities." *Dairy Foods Magazine*, December 1997.

Osterud, Nancy Grey. "'She Helped Me Hay It as Good as a Man': Relations among Women and Men in an Agricultural Community." In *To Toil the Livelong Day: America's Women at Work, 1780–1980*, ed. Carol Groneman and Mary Beth Norton. Ithaca: Cornell University Press, 1987.

Outlaw, Joe L., Robert E. Jacobson, Ronald D. Knutson, and Robert B. Schwart, Jr. *Structure of the U.S. Dairy Farm Sector*. Dairy Markets and Policy: Issues and Options Series, no. M-4. Ithaca: Cornell University Program on Dairy Markets and Policy, 1996.

Paley, William. 1803. *Natural Theology, or, Evidences of the Existence and Attributes of the Deity Collected from the Appearances of Nature*. Albany, NY: Daniel and Samuel Whiting, 1802.

Park, Robert. "Human Migration and the Marginal Man." *American Journal of Sociology* 33 (6) (1928): 881–92.

Park, William H. "The Relation of Milk to Public Health." In New York City Department of Health Milk Commission, *Is Loose Milk a Health Hazard?* New York: Department of Health, 1931.

Parks, Ellen Sorge. "Experiment in the Democratic Planning of Public Agricultural Activity." Ph.D. diss., Department of Political Science, University of Wisconsin-Madison, 1947.

Parsons, Merton S. *Effect of Changes in Milk and Feed Prices and in Other Factors upon Milk Production in New York*. Bulletin 688. Ithaca: Cornell University Agricultural Experiment Station, 1938.

Pegram, Thomas R. *Battling Demon Rum*. Chicago: Ivan R. Dee, 1998.

———. "Public Health and Progressive Dairying in Illinois." *Agricultural History* 65 (1) (1991): 36–50.

Peluso, Nancy Lee. "Coercing Conservation: The Politics of State Resource Control." *Global Environmental Change* 3 (2) (1993): 199–217.

Perrow, Charles. *Normal Accidents: Living with High-Risk Technologies*. New York: Basic Books, 1984.

Petty, Celia. "Food, Poverty and Growth: The Application of Nutrition Science, 1918–1939." *Society for the Social History of Medicine Bulletin* 40 (1987): 37–40.

Pillsbury, Richard. *No Foreign Food: The American Diet in Time and Place*. Boulder: Westview, 1998.

Piore, Michael J., and Charles F. Sabel. *The Second Industrial Divide: Possibilities for Prosperity*. New York: Basic Books, 1984.

Pirtle, Thomas Ross. *History of the Dairy Industry*. Chicago: Mojonnier Brothers, 1926.

Plante, Ellen M. *Women at Home in Victorian America: A Social History*. New York: Facts on File, 1997.

Pollard, A. J., and L. F. Champlin. *Receipts of Milk and Cream at the New York Market*. Unnumbered publication. Washington, DC: USDA, Bureau of Agricultural Economics, 1939.

Popkin, Barry M., T. Lasky, J. Litvin, D. Spicer, and M. E. Yamamoto. *The Infant-Feeding Triad: Infant, Mother, and Household*. Food and Nutrition in History and Anthropology, no. 5. New York: Gordon and Breach, 1986.

Porter, Michael E. *The Competitive Advantage of Nations*. New York: Basic Books, 1990.

Presbrey, F. *The History and Development of Advertising*. New York: Greenwood, 1968 [1929].

Preston, Samuel H., and Michael R. Haines. *Fatal Years: Child Mortality in Late Nineteenth-Century America*. Princeton: Princeton University Press, 1991.

Prout, William. *Chemistry, Meteorology, and the Function of Digestion: Considered with a Reference to Natural Theology*. London: William Pickering, 1834.

Rabinbach, Anson. *The Human Motor: Energy, Fatigue, and the Origins of Modernity*. New York: Basic Books, 1990.

Ray, John. *The Wisdom of God Manifested in the Works of the Creation*. New York: Garland, 1979 [1691].

Raynolds, Laura. "The Restructuring of Export Agriculture in the Dominican Republic." In *The Global Restructuring of Agro-Food Systems*, ed. Philip McMichael. Ithaca: Cornell University Press, 1994.

Redclift, Michael. *Wasted: Counting the Costs of Global Consumption*. London: Earthscan, 1996.

Report of the Joint Legislative Committee to Investigate the Milk Industry. Testimony of George F. Warren. Albany: State of New York, 1933.

Riley, Glenda. *Inventing the American Woman: A Perspective on Women's History*. Arlington Heights, IL: Harlan Davidson, 1986.

Ritter, Gretchen. *Goldbugs and Greenbacks: The Antimonopoly Tradition and the Politics of Finance in America, 1865–1896*. Cambridge: Cambridge University Press, 1997.

Roadhouse, Chester, and James Lloyd Henderson. *The Market-Milk Industry*. 2d ed. New York: McGraw-Hill, 1950.

Rogers, Everett M., and F. Floyd Shoemaker. *Communication of Innovations; A Cross-Cultural Approach*, 2d ed. New York: Free Press, 1971.

Rosaldo, Renato. *Culture and Truth: The Remaking of Social Analysis*. Boston: Beacon, 1989.

Rosenau, M. J. *The Milk Question*. Boston: Houghton Mifflin, 1912.

Rosenberg, Charles, and Carroll Smith-Rosenberg. "Pietism and the Origins of the American Public Health Movement: A Note on John H. Griscom and Robert M. Hartley." *Journal of the History of Medicine* 23 (1968): 16–35.

Rosenberg, Margit. "Breast-Feeding and Infant Mortality in Norway, 1860–1930." *Journal of Biosocial Science* 21 (3) (1989): 335–48.

Rosenkrantz, Barbara. "The Trouble with Bovine Tuberculosis." *Bulletin of the History of Medicine* 59 (2)(1985): 155–75.

Ross, Edward Alsworth. *Standing Room Only?* New York: Century, 1927.

Ross, Ellen. *Love and Toil: Motherhood in Outcast London, 1870–1918*. New York: Oxford University Press, 1993.

Ross, H. A. *The Demand Side of the New York Milk Market*. Bulletin 459. Ithaca: Cornell University Agricultural Experiment Station, 1927.

Rousseau, Jean-Jacques. *Emile*. New York: Dutton, 1911.

Ruddick, Sara. "Thinking Mothers/Conceiving Birth." In *Representations of Motherhood*, ed. Donna Bassin, Margaret Honey, and Meryle Mahrer Kaplan. New Haven: Yale University Press, 1994.

Rush, Benjamin. *An Inquiry into the Effects of Ardent Spirits upon the Human Body and Mind*. 8th ed. Brookfield, MA: E. Merriam, 1814.

Ryan, Mary P. *Cradle of the Middle Class: The Family in Oneida County, New York, 1790–1865*. New York: Cambridge University Press, 1981.

Salamon, Sonya. *Prairie Patrimony: Family, Farming and Community in the Midwest*. Chapel Hill: University of North Carolina Press, 1992.

———. "Ethnic Communities and the Structure of Agriculture." *Rural Sociology* 50 (3) (1985): 323–40.

Salamon, Sonya, and Shirley O'Reilly. "Family Land and Development Cycles among Illinois Farmers." *Rural Sociology* 45 (2) (1979): 290–308.

Salmon, Marylynn. "The Cultural Significance of Breastfeeding and Infant Care in Early Modern England and America." *Journal of Social History* 28 (2) (1994): 247–70.

Salomonsson, Anders. "Milk and Folk Belief: With Examples from Sweden." In *Milk and Milk Products: From Medieval to Modern Times*, ed. Patricia Lysaght. Edinburgh: Canongate Academic, 1994.

Saloutos, Theodore, and John D. Hicks. *Twentieth Century Populism: Agricultural Discontent in the Middle West, 1900–1939*. Lincoln: University of Nebraska Press, 1951.

Sanders, Elizabeth. *Roots of Reform*. Chicago: University of Chicago Press, 1999.

Saum, Lewis O. *The Popular Mood of Pre–Civil War America*. Westport, CT: Greenwood Press, 1980.

Saxe, Stephen O. *Old-Time Advertising Cuts and Typography: 184 Plates from the Boston Type and Stereotype Foundry Catalog (1832)*. New York: Dover, 1989.

Schlebecker, John T. "The World Metropolis and the History of American Agriculture." In *The Use of the Land; Essays on the History of American Agriculture*, ed. John T. Schlebecker. Lawrence, Kan.: Coronado Press, 1973.

Schaars, Marna A. Interview by Donna Taylor. University Archives Oral History Project. Madison: University of Wisconsin, 1978.

Schlossman, Steven L. "The 'Culture of Poverty' in Ante-Bellum Social Thought." *Science and Society* 38 (2) (summer 1974): 150–66.

Schmidt, Louis Bernard, and Earle Dudley Ross. *Readings in the Economic History of American Agriculture*. New York: Macmillan, 1966 [1925].

Scott, Mary. "Organic Dairy a Cash Cow." *Natural Foods Merchandiser*, June 1997.

Shaftel, Norman. "A History of the Purification of Milk in New York; or, 'How Now Brown Cow.'" In *Sickness and Health in America: Readings in the History of Medicine and Public Health*, ed. Judith Walzer Leavitt and Ronald L. Numbers. Madison: University of Wisconsin Press, 1978.

Shover, John L. *First Majority, Last Minority: The Transforming of Rural Life in America*. DeKalb: Northern Illinois University Press, 1976.

Simmonds, N. W. *The State of the Art of Farming Systems Research*. Washington, DC: World Bank, 1984.

Simoons, Frederick J. "The Determinants of Dairying and Milk Use in the Old World: Ecological, Physiological, and Cultural." *Ecology of Food and Nutrition* 2 (1973): 83–90.

Slater, Peter. *Children in the New England Mind: In Death and in Life*. Hamden, CT: Archon Books, 1977.

Slichter, Gertrude Almy. "Franklin D. Roosevelt's Farm Policy as Governor of New York State, 1928–1932." *Agricultural History* 33 (8) (1959): 167–76.

Slovic, Paul. "Public Perception of Risk." *Journal of Environmental Health* 59 (9) (1997): 22–25.

———. "Perceptions of Risk: Reflections on the Psychometric Paradigm." In *Social Theories of Risk*, ed. Sheldon Krimsky and Dominic Golding. Westport, CT: Praeger, 1992.

Smith, Adam. *The Wealth of Nations*. New York: Modern Library, 1973.

Smith-Rosenberg, Carroll. "Beauty, the Beast and the Militant Woman: A Case Study in Sex Roles and Social Stress in Jacksonian America." *American Quarterly* 23 (4) (1971): 562–84.

———. *Religion and the Rise of the American City*. Ithaca: Cornell University Press, 1971.

Spann, Edward K. *The New Metropolis: New York City, 1840–1857*. New York: Columbia University Press, 1981.

Spencer, Leland. *Cooperative Marketing of Milk in New York*. Mimeo L8-1, revised May 31. Ithaca: Cornell University Department of Agricultural Economics, 1932.

Spencer, Leland, and Charles J. Blanford. *An Economic History of Milk Marketing and Pricing, 1800–1933*. Vol. 1. Amended version. Columbus, OH: Grid, 1977.

Spencer, Leland, and S. Kent Christensen. *Milk Control Programs of the Northeastern States*. Part 2. *Administrative and Legal Aspects and Co-ordination of State and Federal Regulation*. Bulletin 918. Ithaca: Cornell University Agricultural Experiment Station, 1955.

Spencer, Leland, and Ida A. Parker. *Consumption of Milk and Cream in the New York City Market and Northern New Jersey*. Bulletin 965. Ithaca: Cornell University Agricultural Experiment Station, 1961.

Stansell, Christine. *City of Women: Sex and Class in New York, 1789–1860*. Urbana: University of Illinois Press, 1987.

Steckel, Lee M. *The Milk Supply of New York City—A Lesson in Municipal Legislation*. New York, 1916.

Steenbock, Harry, and E. B. Hart. *Milk, the Best Food*. Madison: University of Wisconsin Agricultural Experiment Station, 1922.

Stephens, P. H. *Economic Studies of Dairy Farming in New York*, part 11, *Success in Management of Dairy Farms as Affected by the Proportion of the Factors of Production*. Bulletin 562. Ithaca, NY: Cornell University Agricultural Experiment Station, 1933.

Stevens Heininger, Mary Lynn, et al. *Century of Childhood, 1820–1920*. Rochester, NY: Margaret Woodbury Strong Museum, 1984.

Stimpson, Catharine, ed. *Women and the American City*. Chicago: University of Chicago Press, 1981.

Stipp, David. "Is Monsanto's Biotech Worth Less Than a Hill of Beans?" *Fortune*, February 21, 2000, 157–60.

Stonecash, Jeffrey. "Political Parties and Legislative Behavior in New York." Prepared for use by the Assembly Intern Program, New York State Legislature, 1986.

Straus, Nathan. *Disease in Milk: The Remedy Pasteurization: The Life Work of Nathan Straus*. Compiled by Lina Gutherz Straus. New York: Arno Press, 1977 [1917].

Stuart-Macadam, Patricia, and Katherine A. Dettwyler, eds. *Breastfeeding: Biocultural Perspectives*. New York: Aldine de Gruyter, 1995.

Sussman, George D. *Selling Mothers' Milk: The Wet-Nursing Business in France, 1715–1914*. Urbana: University of Illinois Press, 1982.

Swedlund, A. C. "Infant Mortality in Massachusetts and the U.S. in the Nineteenth Century." In *Disease in Populations in Transition: Anthropological and Epidemiological Perspectives*, ed. A. C. Swedlund and G. H. Armelagos. New York: Bergin and Garvey, 1990.

Szasz, Andrew. *Inverted Quarantine*. Minneapolis: University of Minnesota Press, forthcoming.

Tauer, Janelle R., and Olan D. Forker. *Dairy Promotion in the United States, 1979–1986: The History and Structure of the National Milk and Dairy Product Promotion Program with Special Reference to New York*. A.E. 87–5. Ithaca: Cornell University Agricultural Experiment Station, 1987.

Taylor, Henry C. *The History of Agricultural Economics in the United States*. Part 5, *A Farm Economist in Washington, 1919–1925*. Manuscript in Mann Library Collection, Cornell University.

Taylor, Henry C., and Anne Dewees Taylor. *The Story of Agricultural Economics in the United States, 1840–1932*. Vol. 1. Amended version. Columbus, OH: Grid, 1953.

Taylor, Joe Gray. *Eating, Drinking, and Visiting in the South: An Informal History*. Baton Rouge: Louisiana State University Press, 1982.

Terrie, Philip G. *Forever Wild: Environmental Aesthetics and the Adirondack Forest Preserve*. Philadelphia: Temple University Press, 1985.

Teuteberg, Hans J. "The Beginnings of the Modern Milk Age in Germany." In *Food in Perspective*, ed. Alexander Fenton and Trefor M. Owen. Edinburgh: John Donald, 1977.

Thevenot, Laurent. "Innovating in 'Qualified' Markets: Quality, Norms and Conventions." Memorandum to the "Workshop on Systems and Trajectories for Agricultural Innovation," Berkeley, California, April 23–25, 1998.

Thomas, John L. "Romantic Reform in America, 1815–1865." *American Quarterly* 17 (4) (1965): 656–81.

Thornton, P. A., and S. Olson. "Family Contexts of Fertility and Infant Survival in Nineteenth Century Montreal." *Journal of Family History* 16 (1991): 401–17.

Treckel, Paula A. "Breast Feeding and Maternal Sexuality in Colonial America." *Journal of Interdisciplinary History* 20 (1) (1989): 25–51.

Turbin, Carole. "Beyond Conventional Wisdom: Women's Wage Work, Household Economic Contribution, and Labor Activism in a Mid-Nineteenth Century Working Class Community." In *To Toil the Livelong Day: America's Women at Work, 1780–1980,* ed. Carole Groneman and Mary Beth Norton. Ithaca: Cornell University Press, 1987.

Turner, Frederick Jackson. "Sections and Nation." *Yale Review* 12 (1923): 1–21.

Ulrich, Laurel Thatcher. *Good Wives: Image and Reality in the Lives of Women in Northern New England, 1650–1750*. New York: Knopf, 1982.

———. "Housewife and Gadder: Themes of Self-sufficiency and Community in Eighteenth-century New England." In *"To Toil the Livelong Day": America's Women at Work, 1780–1980,* eds. Carol Groneman and Mary Beth Norton. Ithaca, NY: Cornell University Press, 1987.

University of Wisconsin. Extension Service of the College of Agriculture. "Outline for County Land Use Planning in Wisconsin." Mimeo. Madison, 1939.

U.S. Bureau of the Census. Census of Agriculture. Washington, DC: GPO, 1987.

U.S. Department of Agriculture/U.S. Department of Health and Human Services. *Nutrition and Your Health: Dietary Guidelines for Americans,* 5th ed. Washington, DC: GPO, 2000.

U.S. Department of Commerce, Bureau of Economic Analysis. *Survey of Current Business*. Selected issues.

Vandergeest, Peter, and Nancy Peluso. "Territorialization and State Power in Thailand." *Theory and Society* 24 (3) (1995): 385–426.

Van der Ploeg, Jan Douwe. "Styles of Farming: An Introductory Note on Concepts and Methodology." In *Born from Within: Practice and Perspectives of Endogenous Rural Development*, ed. Jan Douwe van der Ploeg and Ann Long. Assen, Netherlands: Van Gorcum, 1994.

Van Winter, Johanna Marie. "The Consumption of Dairy Products in the Netherlands in the 15th and 16th Centuries." In *Milk and Milk Products: From Medieval to Modern Times*, ed. Patricia Lysaght. Edinburgh: Canongate Academic, 1994.

Visness, Cynthia M., and Kathy I. Kennedy. "Maternal Employment and Breast-Feeding: Findings from the 1988 National Maternal and Infant Health Survey." *American Journal of Public Health* 87 (6) (1997): 945–50.

Vitzhum, J. H., and G. R. Benteley. "The Ecology of Breastfeeding." *American Journal of Human Biology* 8 (1996): 102–3.

Von Thünen, Johann Heinrich. *The Isolated State*. New York: Pergamon, 1966.

Walters, Ronald G. *American Reformers, 1815–1860*. New York: Farrar, Straus and Giroux, 1978.

Warren, George F. "A State Program of Agricultural Development." *Journal of Farm Economics* 12 (3) (1930): 359–66.

Warren, Stanley W. *Factors for Success on Dairy and General Farms in Northern Livingston County, New York*. Bulletin 242. Ithaca: Cornell University Cooperative Extension, 1932.

Wasserman, Manfred J. "Henry L. Coit and the Certified Milk Movement in the Development of Modern Pediatrics." *Bulletin of the History of Medicine* 46 (4) (1972): 359–90.

Weber, Max. "Power, Domination, and Legitimacy." In *Power in Modern Societies*, ed. Marvin E. Olsen and Martin N. Marger. Boulder: Westview, 1993.

Weeks, David. *Land Utilization in the Northern Sierra Nevada: Results of a Coöperative Investigation Conducted by the California Agricultural Experiment Station and the California Forest and Range Experiment Station of the Forest Service*. United States Department of Agriculture. Giannini Foundation of Agricultural Economics Contributions 106, University of California at Berkeley, 1943.

Wermuth, Thomas S. "New York Farmers and the Market Revolution: Economic Behavior in the Mid-Hudson Valley, 1780–1830." *Journal of Social History* 32 (1) (1998): 179–97.

Whatmore, Sarah, and Lorraine Thorne. "Nourishing Networks: Alternative Geographies of Food." In *Globalizing Food: Agrarian Questions and Global Restructuring*, ed. David Goodman and Michael Watts. New York: Routledge, 1997.

Whiteaker, Larry. *Seduction, Prostitution, and Moral Reform in New York, 1830–1860*. New York: Garland, 1997.

Wiebe, Robert. *The Search for Order, 1877–1920*. New York: Farrar, Straus and Giroux, 1967.

Wiles, Richard, Kert Davies, and Susan Elderkin. "A Shopper's Guide to Pesticides in Produce." *Environmental Working Group*, November 1995.

Willard, Xerxes. "American Dairying: Its Rise, Progress, and National Importance." In *United States Department of Agriculture Report*. Washington, DC: GPO, 1866.

Williams, Naomi, and Chris Galley. "Urban-Rural Differentials in Infant Mortality in Victorian England." *Population Studies* 49 (1995): 401–20.

Williams, Sheldon W., David A. Vose, Charles E. French, Hugh L. Cook, and Alden C. Manchester. *Organization and Competition in the Midwest Dairy Industry*. Ames: Iowa State University Press, 1970.

Wisconsin Statistical Reporting Service. *Wisconsin Dairy Herds by Type of Milk Produced, Number and Percent by County*. Madison, 1976.

Wolcott, Leon. "National Land-Use Programs and the Local Governments." *National Municipal Review* 29 (1939): 111–19.

Wolf, Jacqueline H. "Don't Kill Your Baby: Feeding Infants in Chicago, 1903–1924." *Journal of the History of Medicine* 53 (1998): 219–53.

Woodbury, Robert Morse. *Infant Mortality and Its Causes.* Baltimore: Williams and Wilkins, 1926.

Woods, Michael. "Rethinking Elites: Networks, Space, and Local Politics." *Environment and Planning* A 30 (12) (1998): 2101–19.

Wooten, H. H. *The Land Utilization Program, 1934 to 1964: Origin, Development and Present Status.* Agricultural Economic Report 85. USDA, Economic Research Service. Washington, DC: GPO, 1965.

Worster, Donald. *Nature's Economy: A History of Ecological Ideas.* New York: Cambridge University Press, 1977.

Wyngate, Pamela. "Organic Dairy: The Little Niche That Could."*Natural Foods Merchandiser*, May 1999.

Young, Brigitte. "Does the American Dairy Industry Fit the Meso-Corporatist Model?" *Political Studies* 38 (1) (1990): 72–82.

Zuckerman, Larry. *The Potato: How the Humble Spud Rescued the Western World.* Boston: Faber and Faber, 1998.

Index

Numbers in italics indicate figures and tables.

About the Author

E. MELANIE DUPUIS is an assistant professor in the Department of Sociology at the University of California, Santa Cruz. Previously, she worked as a policy analyst for the state of New York. She is the coeditor of *Creating the Countryside: The Politics of Rural and Environmental Discourse.*